REPORT

Commercial Intratheater Airlift

Cost-Effectiveness Analysis of Use in U.S. Central Command

Ronald G. McGarvey • Thomas Light • Brent Thomas • Ricardo Sanchez

Prepared for the United States Air Force
Approved for public release; distribution unlimited

The research described in this report was sponsored by the United States Air Force under Contract FA7014-06-C-0001. Further information may be obtained from the Strategic Planning Division, Directorate of Plans, Hq USAF.

Library of Congress Cataloging-in-Publication Data

McGarvey, Ronald G.
Commercial intratheater airlift : cost-effectiveness analysis of use in U.S. Central Command / Ronald G. McGarvey, Thomas Light, Brent Thomas, Ricardo Sanchez.
pages cm
Includes bibliographical references.
ISBN 978-0-8330-7837-7 (pbk. : alk. paper)
1. Airlift, Military—United States—Costs—Evaluation. 2. Airlift, Military—Contracting out—United States—Evaluation. 3. United States. Central Command. I. Light, Thomas, Ph. D. II. Thomas, Brent. III. Sanchez, Ricardo R., 1979- IV. Title. V. Title: CITA, cost-effectiveness analysis of use in US CENTCOM.

UC333.M35 2013
358.4'40681—dc23

2013004955

The RAND Corporation is a nonprofit institution that helps improve policy and decisionmaking through research and analysis. RAND's publications do not necessarily reflect the opinions of its research clients and sponsors.

Published 2013 by the RAND Corporation
1776 Main Street, P.O. Box 2138, Santa Monica, CA 90407-2138
1200 South Hayes Street, Arlington, VA 22202-5050
4570 Fifth Avenue, Suite 600, Pittsburgh, PA 15213-2665
RAND URL: http://www.rand.org/
To order RAND documents or to obtain additional information, contact
Distribution Services: Telephone: (310) 451-7002;
Fax: (310) 451-6915; Email: order@rand.org

Preface

The Department of Defense (DoD) has spent hundreds of millions of dollars on commercial intratheater airlift (CITA) movements in the U.S. Central Command (USCENTCOM) area of responsibility. This is notable, beyond simply the magnitude of the expenditures, because intratheater airlift within a combat theater of operation is typically assumed to be a mission performed by military aircraft; indeed, the U.S. Air Force (USAF) has deployed a number of C-130s and C-17s to USCENTCOM in support of this specific mission.

Although some of the motivations provided for the use of CITA in this particular case (that is, within USCENTCOM) are inherently noneconomic, the analysis detailed in this report aims to answer the following question: Were these expenditures on CITA cost-effective, relative to the cost of performing these same movements on USAF-organic aircraft? That is, did DoD get a "good value" on these purchases? To answer that question, we first had to identify how to measure cost-effectiveness of CITA. We then examined whether the use of CITA should have been expanded or reduced, relative to the historical experience.

The research described in this report was conducted within the Resource Management Program of RAND Project AIR FORCE for a project titled "A Civil Reserve Air Fleet (CRAF)–Like Concept for Intratheater Lift: Approaches and Effects on Requirements," sponsored by Maj Gen Brooks Bash, former Director of Operations, Headquarters Air Mobility Command.

This report should be of interest to mobility planners, logisticians, and contracting personnel throughout DoD, in particular those associated with U.S. Transportation Command.

RAND Project AIR FORCE

Contents

APPENDIXES

Figures

Tables

Summary

Intratheater airlift (ITA) is used to deliver critical and time-sensitive supplies, such as blood products for transfusions or repair parts for vehicles, to deployed forces. ITA within a combat theater of operation has traditionally been assumed to be provided by military aircraft. However, in recent years, a number of commercial providers have been providing a significant amount of ITA within U.S. Central Command (USCENTCOM). A number of motivations for the use of commercial intratheater airlift (CITA) within USCENTCOM have been identified, such as[1]

- concerns about structural fatigue to C-130 aircraft due to the heavy use of these aircraft in USCENTCOM
- lack of access to C-130 aircrews, particularly those in the Air Reserve Component, for deployments
- a desire to reduce the use of convoys over high-threat roadways.

The analysis detailed in this report aims to answer the following question: Were these expenditures on CITA cost-effective, relative to the cost of performing these same movements on organic U.S. Air Force (USAF) aircraft? That is, did the Department of Defense (DoD) get a "good value" on these purchases?

When multiple airlift options exist for any specific movement, it is first necessary to identify the cost to provide each movement via each airlift option. Identifying such a cost is complicated by two primary factors: for commercial carriers, the extent of *price elasticities of demand* (to what extent do changes in the demand for CITA movements impact the price charged by commercial carriers for these movements?); for USAF-organic aircraft, the set of *which fixed and marginal costs to include* (is the procurement or retirement of aircraft under consideration, or is one only considering variations to the number of flying hours performed by aircraft that are already in the USAF inventory, and whose inventory levels are justified according to some requirement exogenous to their use in this specific scenario?).

Moreover, identifying the cost for each individual movement does not capture the potential for reducing costs by *aggregating multiple movements across aircraft sorties and missions*. Given a large collection of movement requirements and a set of airlift alternatives, it is nec-

[1] Another potential motivation that has been suggested, beyond the three listed here, is a desire to foster the development of more-robust commercial logistic options in the region to support reconstruction and other local, nonmilitary development activities. Given that the carriers providing CITA were not indigenous Iraqi or Afghan firms and that, by the nature of the cargo airlift industry, these aircraft will move around the globe in response to the demands for their services, we find this motivation to be somewhat lacking in validity.

essary to solve a routing problem and an assignment problem: which movements to assign to which missions. We developed an optimization model to identify the most cost-effective movement-to-mission assignments for any set of movement requirements and available airlift alternatives.

Conclusions

Based on the demands for ITA and the costs paid to CITA providers in USCENTCOM in 2009, we find:

C-17 and C-130 are both generally more cost-effective than CITA, but CITA options should be retained to supplement USAF aircraft. Across all our optimization model runs, the model demonstrated a clear preference for the increased use of USAF aircraft and the decreased use of CITA, although, for a relatively small fraction of the USCENTCOM ITA demands, the IL-76 charters and Theater Express Program (TEP) tenders were the most cost-effective options.

As shown in Figure S.1, for a level of USAF resources equal to that used in USCENTCOM (15 C-17s and 21 C-130s available each day for ITA movements), the optimized allocation of cargo and passengers to airlift options reduced costs by $175 million from the historical performance. The model achieved these savings by replacing the TEP tenders with C-17s for long movements and by replacing TEP with IL-76 charters for short movements. The optimization model made these changes primarily because of the ability of the C-17 and IL-76 to aggregate cargo across multisortie missions, supporting a mix of high-demand and low-demand origin-destination pairs on a single mission. Using TEP to provide the same sets

Figure S.1
Total Delivery Cost for Optimization Model Results and Experience

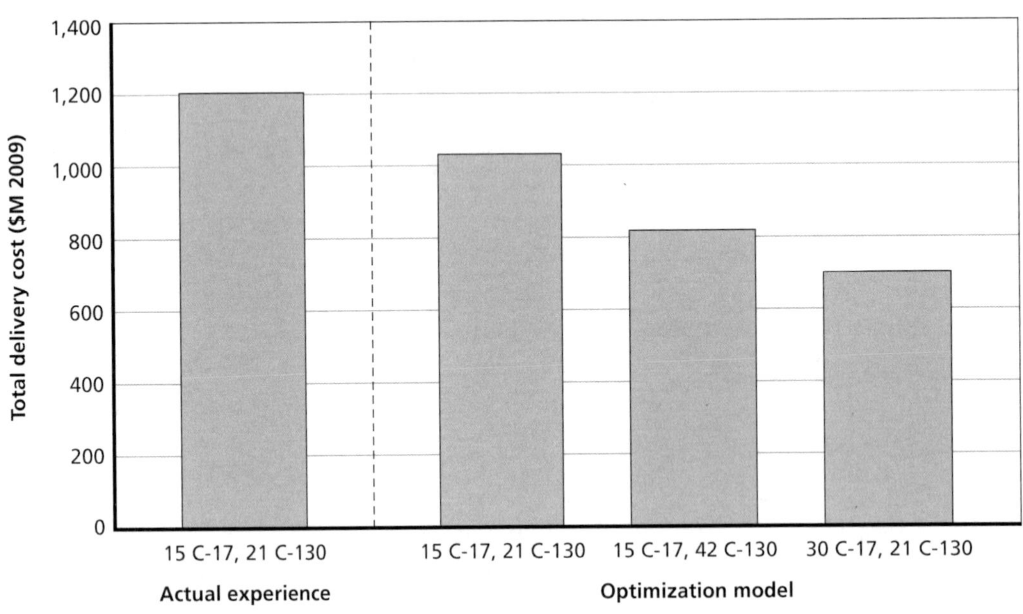

of movements would require a separate tender for each origin-destination pair, with the low-demand origin-destination pairs driving a comparatively high cost per amount moved.[2]

If the level of C-130s supporting USCENTCOM ITA could be doubled, the optimized allocation could reduce costs by slightly more than $210 million over the optimization result for 21 C-130s (for a total savings of approximately $390 million). These $210 million savings accrue because the increased number of C-130s allows significantly more passengers movements to occur on C-130s, with a corresponding decrease in C-17 passenger movements, thereby freeing C-17s to replace IL-76s for long cargo movements (some of the additional C-130s are also used to replace IL-76s for short cargo movements). Doubling the level of C-17s supporting USCENTCOM ITA while holding C-130s at the historical level could allow the optimized allocation to reduce costs by slightly more than $330 million over the optimization result for 15 C-17s (for a total savings of approximately $510 million). These $330 million savings are accrued by significantly increasing both the passenger and cargo movements occurring on C-17s, with a corresponding decrease in C-130 passenger movements and cargo movements using TEP and IL-76 charters. In both cases, the increased usage of C-17s for cargo movements is due to the C-17's relative cost advantage: On a cost-per-mile basis, the IL-76 costs three times as much as the C-17, for a comparable aircraft block speed and payload. These cost reductions do not account for the increased costs to the USAF associated with deploying additional aircraft to USCENTCOM, but our preliminary analysis suggests that these deployment costs would be much smaller than the potential savings.

The minimum cost that was achieved with 30 C-17s was essentially equal to the global minimum cost achievable if we allowed the model access to an unlimited pool of C-130s and C-17s. Because the model's solution with 30 C-17s utilized TEP and IL-76s to transport 4 and 6 percent, respectively, of the total cargo tonnage, CITA appears to be the most cost-effective delivery option for these cargoes, suggesting that, even though USAF aircraft appear to be more cost-effective for most of the USCENTCOM ITA movements, CITA options should be retained for some small fraction of movements.[3]

We also performed sensitivity analyses, running the optimization model with lower per-movement costs for TEP and higher costs for C-130s, C-17s, and IL-76s. In these sensitivity analyses, the amount of cargo transported via C-17 and C-130 did not significantly change, suggesting that our finding that USAF aircraft are more cost-effective than commercial alternatives for most ITA movements within USCENTCOM is fairly robust. However, as IL-76 costs increased and TEP costs decreased, the amount of cargo transported via TEP increased, with most of this increase coming at the expense of cargo previously transported via IL-76. This suggests that, while IL-76s may be slightly more cost-effective than TEP for most of the ITA movements within USCENTCOM best suited for CITA, this preference is not robust to moderately sized changes to the relative IL-76 and TEP cost structures.

Decision-support tools are needed to assist the Combined Air and Space Operations Center (CAOC) Air Mobility Division (AMD) and USCENTCOM Deployment and Distribution Operations Center with daily airlift cargo allocation decisions. Within

---

[2] Note that, in other applications, such as contract trucking movements in the United States, the use of *combinatorial auctions* allows competitors to submit bids on any desired combination of individual movement requirements; optimization routines can be used to select the set of bids that spans the requirements at minimum cost. Our analysis did not examine the potential benefits of such an application to the TEP; rather, we examined the TEP as it existed in 2009.

[3] This assumes that CITA providers would still elect to participate at the reduced volumes.

USCENTCOM, each day the CAOC AMD, based on guidance from the USCENTCOM Deployment and Distribution Operations Center, must assign movement requirements to airlift options. Given a large collection of movement requirements and a set of airlift alternatives, this entails solving a routing problem and an assignment problem: which movements to assign to which missions. At the time of this analysis, the CAOC AMD did not have access to sophisticated decision-support tools to assist in its daily determination of cargo-aircraft allocation decisions. The extremely large number of potential assignments prohibits any individual from considering all feasible options and selecting the most effective solution without the aid of a computer model. We developed an optimization model to perform such movement-to-mission assignments and found that the model was able to identify significant improvements to the historical performance. As noted in Figure S.1, in contrast to USCENTCOM historical performance of approximately $1,210 million in total ITA delivery cost, the model found a solution that could have reduced this cost by up to $175 million, without increasing the number of employed C-17 and C-130 flying each day in theater. This suggests that an investment in the development of such tools for AMD use could achieve large savings.

Acknowledgments

Many people inside and outside the Air Force provided valuable assistance and support to our work.[1] We thank Maj Gen Brooks Bash, Directorate of Operations, Headquarters Air Mobility Command (AMC/A3), and Maj Gen Mark Solo, Commander, 618 Tanker Airlift Control Center (618 TACC/CC), who were the original sponsors of this project, along with Maj Gen Frederick Martin, AMC/A3, who served as our sponsor when this project was completed, for their support. We also thank Maj Gen Robert McMahon, Commander, Warner Robins Air Logistics Center, AMC (WR-ALC/CC), for sharing his feedback on this work, drawing on his recent experiences as the Director of the USCENTCOM Deployment and Distribution Operations Center.

We extend a special thanks to Dave Merrill and Don Anderson in Directorate of Analyses, Assessments and Lessons Learned, Headquarters Air Mobility Command (AMC/A9) for all their assistance throughout this study, from their help with data collection at the outset to their feedback and suggestions as we prepared the final conclusions. The first author adds a further, personal, thanks to Dave Merrill for hosting him in the AMC/A9 office space during his January 2008 to February 2010 assignment at Scott Air Force Base.

A number of other individuals at Scott Air Force Base also assisted our data collection efforts; in particular, we would like to thank Greg Hunt, USTC/AQ and 2nd Lt Chris Jones, AMC/A9, for their efforts to obtain TEP data for our study. Sam Newberry and Sandy Halama at USTC/AQ greatly aided our data collection efforts for the IL-76 and AN-124 aircraft charters in USCENTCOM. Merle Lyman and Mark Caslen at AMC/A3B helped us to better understand the Civil Reserve Air Fleet program. Finally, we thank Bruce Busler, USTC/AC, for taking the general concepts developed in this report and beginning to implement the use of similar decision support tools within USTRANSCOM.

At RAND, we especially thank Mary Chenoweth for her assistance obtaining IL-76 and AN-124 charter aircraft movement data. Anthony Rosello was an extremely useful source of general knowledge about Air Force mobility operations. We also thank General (retired) John Handy, formerly of the RAND Board of Trustees; Michael Kennedy; David Orletsky; Sean Bednarz; Richard Moore; Laura Baldwin; James Masters; and Carl Rhodes for sharing their insights and suggestions during the development of our final project briefing. We would especially like to thank our RAND colleagues Marc Robbins and Christopher Mouton for their thorough reviews; their comments helped shape this monograph into its final, improved form.

[1] All office symbols and military ranks are listed as of the time of this research.

That we received help and insights from those acknowledged above should not be taken to imply that they concur with the findings presented in this report. As always, the analysis and conclusions are solely the responsibility of the authors.

Abbreviations

AFB	Air Force base
AFCAIG	Air Force Cost Analysis Improvement Group
AFDD	Air Force doctrine document
AFI	Air Force instruction
AFMC	Air Force Materiel Command
AFPAM	Air Force pamphlet
AFRC	Air Force Reserve Command
AGR	Active Guard Reserve
ALD	available-to-load date
AMC	Air Mobility Command
AMC/A3	Directorate of Operations, Headquarters Air Mobility Command
AMD	Air Mobility Division
ANG	Air National Guard
AOR	area of responsibility
ARC	Air Reserve Component
ART	Air Reserve Technician
BOS	base operating support
CAOC	Combined Air and Space Operations Center
CITA	commercial intratheater airlift
CORE	Cost Oriented Resource Estimating
CPU	central processing unit
CRAF	Civil Reserve Air Fleet
DoD	Department of Defense
EBH	equivalent baseline hour(s)

FH	flying hour(s)
FY	fiscal year
GATES	Global Air Transportation Execution System
GDSS	Global Decision Support System
ITA	intratheater airlift
MAJCOM	major command
MCRS-16	Mobility Capabilities and Requirements Study 2016
MDS	mission design series
MILP	mixed integer linear program
OMB	Office of Management and Budget
PAA	primary aircraft authorized
PACAF	Pacific Air Forces
PAF	Project AIR FORCE
PAX	passengers
RC	reserve component
RDD	required delivery date
SAAM	special assignment airlift mission
SF	severity factor
STAR	standard theater airlift routes
START	Strategic Tool for the Analysis of Required Transportation
TAI	total aircraft inventory
TCTO	Time Compliance Technical Order
TEP	Theater Express Program
TWCF	Transportation Working Capital Fund
USAF	U.S. Air Force
USAFE	U.S. Air Forces in Europe
USC	U.S. Code
USAFRICOM	U.S. Africa Command
USCENTCOM	U.S. Central Command
USTRANSCOM	U.S. Transportation Command

CHAPTER ONE

Introduction

Since 2001, the Department of Defense (DoD) has sustained a large deployment of military personnel and equipment to the U.S. Central Command (USCENTCOM) area of responsibility (AOR). Much of these deployed forces' demands for critical and time-sensitive supplies, such as blood products for transfusions or repair parts for vehicles, are delivered using intratheater airlift (ITA). By ITA, we are referring to an aircraft sortie whose origin and destination are both in the same geographic region; in this case, the USCENTCOM AOR.

ITA within a combat theater of operation has traditionally been assumed to be provided by military aircraft. Air Force Doctrine Document (AFDD) 2-6, *Air Mobility Operations*, is the keystone reference for U.S. Air Force (USAF) airlift doctrine, presenting the tenets of air mobility for both inter- and intratheater operations. However, while AFDD 2-6 provides explicit guidance for the inclusion of Civil Reserve Air Fleet (CRAF) capabilities into global command structures for intertheater airlift, the sections addressing regional control of ITA do not mention commercial carriers.

Similarly, the Mobility Capabilities and Requirements Study 2016 (MCRS-16) analyzed ITA requirements, using a combination of C-130s, C-17s, and C-27s to provide all ITA movements.[1] MCRS-16 concluded that the

> programmed fleet of 401 C-130s exceeds the peak demand in each of the three MCRS cases. . . . [H]owever, based on current total force planning objectives, the C-130 crew force structure cannot sustain steady state operations in combination with a long duration irregular warfare campaign.

Note this does not mention the use of commercial intratheater airlift (CITA) as a means of alleviating this potential shortfall for sustained support to long-duration deployments.

However, circa 2009, a number of commercial providers were providing a significant share of ITA within USCENTCOM, supporting both very specialized demands, such as delivering fresh fruit and vegetables, and more-standard cargo movements. A number of motivations for the use of CITA within USCENTCOM have been identified, such as[2]

[1] MCRS-16 was intended to "provide an updated, comprehensive assessment of the Department's mobility system, one which could be used to inform the 2010 Quadrennial Defense Review." Although the final MCRS-16 report is classified, an unclassified executive summary was released in February 2010 (DoD, 2010). All references to MCRS-16 contained in this report were obtained from this unclassified executive summary.

[2] Another potential motivation that has been suggested, beyond the three listed here, is a desire to foster the development of more-robust commercial logistic options in the region to support reconstruction and other local, nonmilitary development activities. Given that the carriers providing CITA were not indigenous Iraqi or Afghan firms and that, by the nature

- concerns about the appearance of cracks in C-130 wing-boxes due to the heavy use of these aircraft in USCENTCOM (Anderson, undated)
- lack of access to C-130 aircrews due to "mobilization authority for the Air Reserve Component (ARC) forces expiring" (Omdal, 2010)
- a desire to "reduce the need for convoys on highly trafficked Iraqi roads riddled with improvised explosive devices" (Huard, 2011).

The analysis detailed in this report aims to answer the following question: Were these expenditures on CITA cost-effective, relative to the cost of performing these same movements on USAF-organic aircraft? That is, did DoD get a "good value" on these purchases?

Answering this question requires determining how to measure the cost-effectiveness of providing ITA, and then using that measure to assess whether the use of CITA should have been expanded, or reduced, relative to the historical record.

Of the three motivations for use of CITA presented above, only the first (delay the onset of structural damage to C-130s) can be fully accommodated in a standard cost-effectiveness analysis, since a cost, in dollars, can be identified for the maintenance and replacement of aircraft. Such a cost-effectiveness analysis can address some aspects of the second motivation (access to ARC forces), by identifying the number of additional USAF aircraft that would have been needed in USCENTCOM to execute the movements performed by CITA. Our analysis did not examine the effectiveness of CITA movements at reducing the need for road convoys. However, to the extent that CITA movements were less costly than the USAF-organic alternative, CITA would provide a "win-win" in terms of (necessarily) fewer convoys and reduced costs; were CITA movements more costly, the difference in costs could be viewed as the premium DoD paid to achieve this objective without deploying more USAF aircraft.

The remainder of this report is organized as follows. Chapter Two describes the primary means by which CITA was provided in USCENTCOM, circa 2009, and presents data demonstrating the level of ITA utilization in this theater during this time frame. Chapter Three discusses some of the difficulties associated with determining cost-effectiveness of providing ITA, details the approaches that we developed to generate cost-estimating relationships for ITA movements, and presents the optimization model we developed to identify the minimum cost achievable with a given set of airlift resources. Chapter Four then describes the application of this cost-effectiveness determination procedure to the set of USCENTCOM ITA movements in 2009. Chapter Five concludes the report proper with our findings and recommendations. Appendix A describes the data merging we did, and Appendix B describes our evaluation of the Theater Express Program (TEP). Appendix C shows how we estimated the full marginal costs of using C-130s. Finally, Appendix D describes our CITA optimization model.

---

of the cargo airlift industry, these aircraft will move around the globe in response to the demands for their services, we find this motivation to be somewhat lacking in validity.

CHAPTER TWO

ITA in USCENTCOM

Excluding the specialized movements mentioned in the previous chapter (e.g., deliveries of fresh fruit and vegetables), there are two broad categories of ITA in support of combat operations. *General support* refers to ongoing sustainment of military units, in which requests for airlift are approved and prioritized by a joint force commander. USAF C-130s typically provide such ITA, although C-17s have also provided a significant amount of general support airlift to USCENTCOM, through the Theater Direct Delivery program. *Direct support* refers to immediate support for U.S. Army units, which circumvents the delays associated with the joint approval and prioritization process by allowing the Army to direct the movements of associated aircraft; these movements have historically been performed by Army C-23s, although this mission is being transferred to USAF C-27s as they enter the USAF inventory.[1]

Each day, the Air Mobility Division (AMD) of USCENTCOM's Combined Air and Space Operations Center (CAOC) reviews the set of approved airlift requests and determines whether the movements will occur via commercial or military aircraft, basing the decisions on guidance it receives from the USCENTCOM Deployment and Distribution Operations Center. Because general support accounts for most of the ITA in USCENTCOM, we limit our focus for this analysis to two USAF aircraft, the C-130 and C-17, and to the two means that provided the preponderance of general support CITA: the chartering of large fixed-wing aircraft (IL-76s and AN-124s) and TEP.[2] Thus, we concentrated on the set of movements for which the AMD makes the determination between military or commercial airlift.

Under an aircraft charter, DoD leases the use of an entire aircraft for a specified duration. The U.S. Transportation Command (USTRANSCOM) Acquisition Directorate provided us with copies of the contract language for the IL-76 and AN-124 charter arrangements within USCENTCOM; these contracts specify the rates charged for aircraft missions between various origin-destination pairs and are valid for durations ranging from 30 days to one year. An important characteristic of these contracts is the cancellation fee structure. These contracts guarantee a minimum level of activity to the carriers, generally stated in terms of a minimum number of aircraft missions to be flown on each day. If DoD uses fewer than this minimum number of aircraft missions on a day, it incurs a cancellation penalty; as a result, DoD pays the carrier a fee even though the carrier is not providing any airlift on that day.

[1] After we completed this analysis, the FY 2013 Budget Overview stated that the Air Force plans to "divest the C-27J fleet by retiring 21 aircraft and canceling procurement of 17 additional aircraft" (USAF, 2012).

[2] Although other types of commercial aircraft made ITA deliveries, including rotary-wing aircraft and very small fixed-wing aircraft, such as the CASA C-212, their movements were generally of a nature that would not allow a direct comparison with delivery via USAF C-130s and C-17s (such as deliveries to very short, unimproved airfields).

The IL-76 and AN-124 aircraft that perform the majority of charter movements within USCENTCOM are operated by foreign-flagged carriers working as subcontractors to CRAF members (all of which are U.S.-flagged). Although the Fly America Act (49 U.S. Code [USC] 40118) and Fly CRAF Act (49 USC 41106) require all DoD agencies to use CRAF carriers to transport personnel and cargo if service is reasonably available between two locations outside the United States, U.S.-flagged carriers (and by extension, CRAF members) were not able to fly into many DoD destinations in USCENTCOM because of this theater's threat environment circa 2009, providing an exception-based rationale for the use of these foreign-flagged carriers to perform cargo movements. However, because of the 10 USC 2640 restrictions requiring DoD safety oversight for charter air transportation of members of the armed forces, the IL-76s and AN-124s generally did not move personnel in USCENTCOM, the primary exception being the pallet riders who are required to travel with classified cargo to ensure its control.

TEP is a DoD program that uses commercial carriers to move cargo within USCENTCOM.[3] Each day, the AMD identifies a set of tender movements and offers them to a set of approved carriers. Because TEP is a tendering arrangement, as opposed to a contracted carriage arrangement, these carriers can offer a bid to transport each individual movement, but the carriers are not required to bid on any specific movement, and DoD is not required to offer any specified minimum level of cargo.[4] Each bid received is evaluated based on a combination of the cost offered and the carrier's past performance in delivering awarded cargo within the specified timelines. TEP cannot transport explosives; cargo requiring ventilation (e.g., liquid oxygen carts); wet or dry ice shipments; registered mail; cargo requiring an escort, courier, or signature and tally record; or personnel. A primary distinction between TEP and the whole aircraft charters is that TEP affords DoD much more flexible terms, since no guaranteed minimum level of airlift activity needs to be specified in advance for each day's operations.

Level of ITA Utilization in USCENTCOM

To determine the historical level of CITA use within USCENTCOM, it was necessary to merge cargo and passenger movement data from multiple data systems. The Global Air Transportation Execution System (GATES) collects data on the cargo and passengers that pass through Air Force aerial ports. The Global Decision Support System (GDSS) is a command and control system for the dissemination of airlift and tanker mission plans for all mobility air force operations. GATES and GDSS accumulate data for both USAF-organic aircraft and chartered commercial cargo aircraft controlled by Air Mobility Command (AMC). By linking the cargo and passenger data from GATES with the sortie data from GDSS, we were able to determine the set of all cargo and passengers that moved on at least one intra-USCENTCOM

[3] Circa 2009, TEP also served one location outside of USCENTCOM, Djibouti-Ambouli International Airport, which is in the U.S. Africa Command (USAFRICOM) AOR.

[4] Note that in other applications, such as contract trucking in the United States, the use of *combinatorial auctions* allows competitors to bid on any desired combination of individual movement requirements; optimization routines can be used to select the set of bids that spans the requirements at minimum cost. Our analysis did not examine the potential benefits of such an application to the TEP; rather, we examined the TEP as it existed in 2009.

sortie.[5] Appendix A provides further detail on the process we developed to merge these data sets. The USTRANSCOM Acquisition Directorate provided us with a separate data set that detailed the movements performed by TEP. Combining these data sources, we identified the total cargo tonnage and passenger count for USCENTCOM ITA (for the subset of AMD-assigned movements described above).[6] Although TEP was established in 2006 and although charter aircraft were providing ITA in USCENTCOM prior to that date, we had a complete data set only for calendar year 2009; thus, our analysis is limited to ITA movements during 2009.

As discussed above, USAF C-17s and C-130s move both cargo and passengers in USCENTCOM. Figure 2.1 presents the daily intra-USCENTCOM movements for these aircraft, with the solid lines presenting the tons of cargo moved per day, and the dotted lines presenting the passengers moved per day. The levels of cargo movement were relatively consistent across the year, with the C-17s carrying significantly more cargo than the C-130s; summing across 2009, C-17s transported over three times as many tons of intra-USCENTCOM cargo as did C-130s. However, for passengers, the two aircraft types appear to provide a comparable amount of movement; in fact, summing the passenger movements across the entire year, C-130s performed 52 percent of the total passenger movements.

Figure 2.1
C-17 and C-130 Intra-USCENTCOM Daily Airlift Cargo and Passengers, 2009

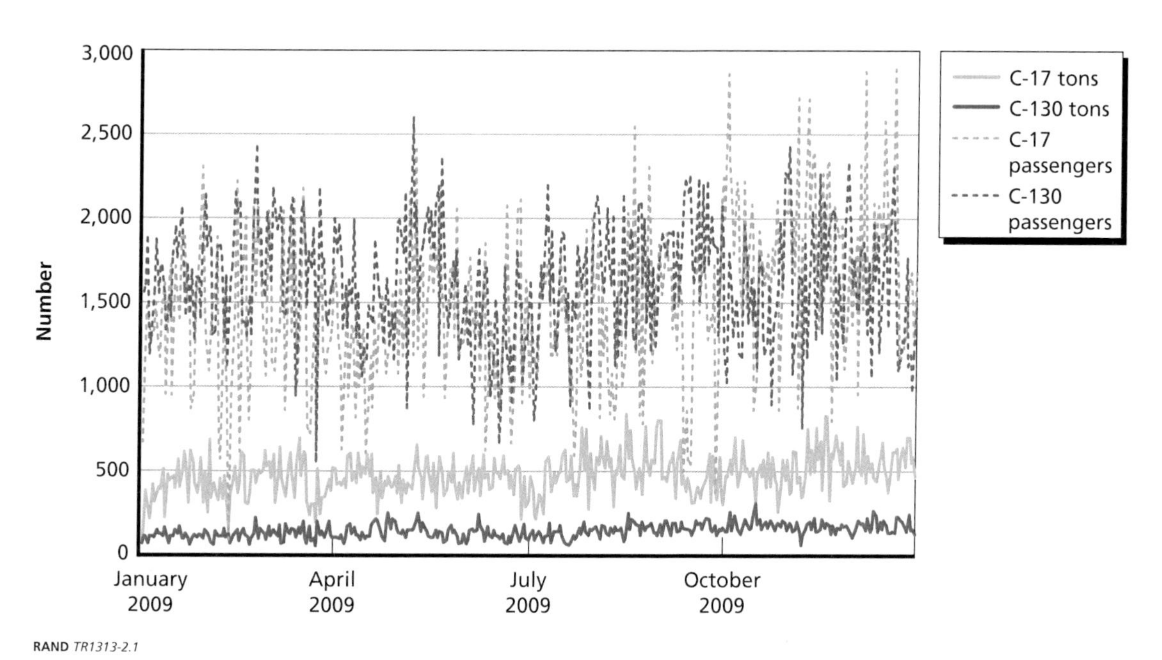

RAND *TR1313-2.1*

[5] Our data set potentially includes some cargo and passengers that were simply transiting USCENTCOM while moving between two other theaters. As a practical matter, this was likely a very rare occurrence and should thus account for very little in the way of cargo or passengers.

[6] The data treatment did not track individual pallets across multiple aircraft; thus, any pallet that was *cross-loaded* (e.g., flying initially on a C-130, then moving to a C-17 to complete the journey) would appear under the movement totals for both C-130s and C-17s. Such cross-loading, while fairly common for cargo moving between theaters, is much less common for ITA movements.

To determine the level of USAF aircraft activity associated with this airlift, we identified the number of daily C-17 and C-130 sorties that were associated with the ITA movements presented in Figure 2.1. Figure 2.2 presents these sortie counts. Across 2009, a relatively consistent number of sorties took place each day for each of these aircraft types, with the C-130 typically performing approximately twice as many daily sorties as the C-17.

We also determined the number of USAF aircraft that were utilized to perform these sorties. Figure 2.3 presents the number of unique C-130 and C-17 tail numbers that flew at least one intra-USCENTCOM sortie each day across the year. While there was some variation, on average, between 14 and 16 unique C-17s flew an intra-USCENTCOM sortie per day. For the C-130s, the average was between 20 and 22 unique aircraft flying such sorties each day. It is important to note here that this number of "employed" aircraft is different from the number of aircraft that were deployed to USCENTCOM during this time frame. Over this interval, an average of 13 C-17s and 29 C-130s were deployed to USCENTCOM locations at any one time (Anderson, 2011). The number of deployed C-17s is less than the number of employed C-17s because C-17s often perform ITA sorties in the course of intertheater missions that do not require aircraft deployment. However, C-130s do not typically provide such intertheater movements. For a C-130 to perform an ITA movement within USCENTCOM, the C-130 would need to be deployed to USCENTCOM. Thus, more deployed C-130s are needed than the daily average of 20 to 22 employed, to account for non–mission capable C-130s and for aircraft that are not tasked to fly on a given day for other reasons.

Although CITA did not move passengers in USCENTCOM, it did move a significant amount of ITA cargo, as shown in Figure 2.4.[7] Observe that over 40 percent of the total 2009

Figure 2.2
C-17 and C-130 Intra-USCENTCOM Daily Airlift Sorties, 2009

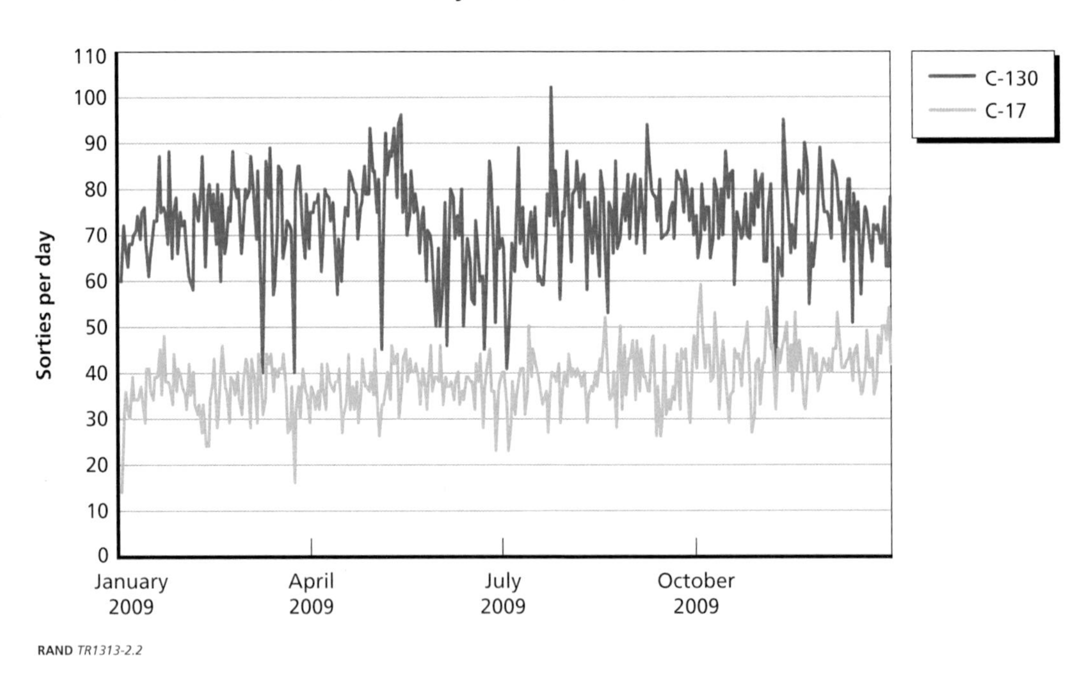

RAND *TR1313-2.2*

[7] Because our TEP data set did not include movements for the last two weeks of 2009, we based our data set's tender movements for December 18–31, 2009, on the TEP movements performed during December 18–31, 2008 (for which we had data).

Figure 2.3
Numbers of C-17 and C-130 Aircraft Flying at Least One Intra-USCENTCOM Sortie per Day, 2009

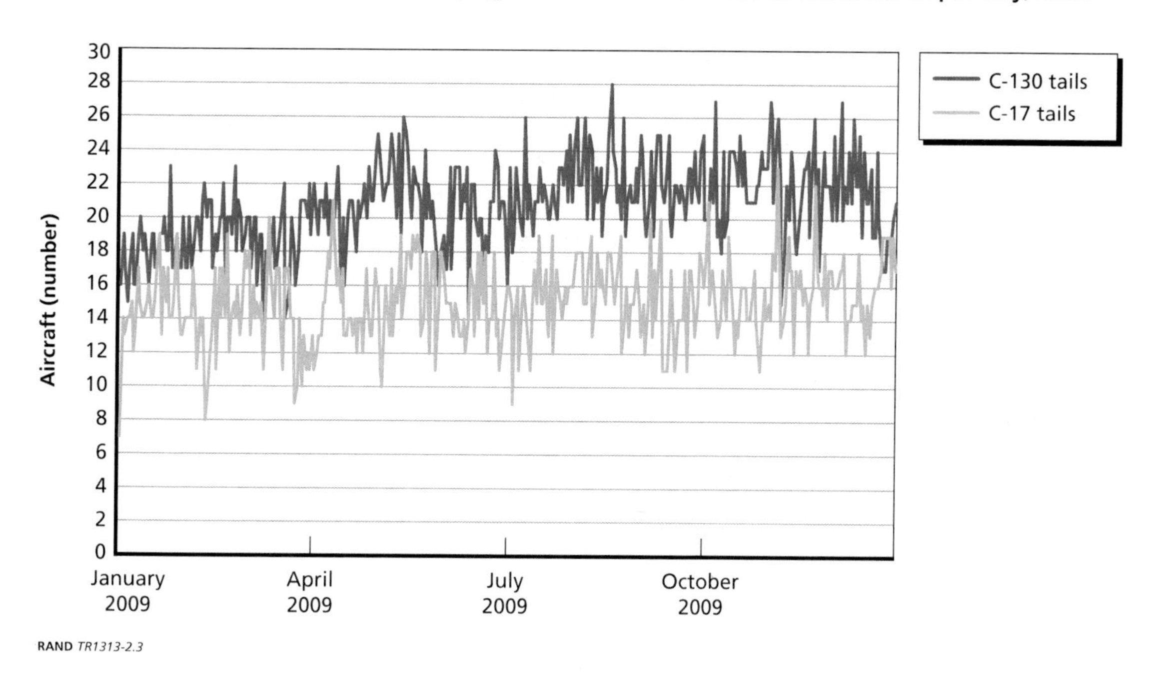

RAND *TR1313-2.3*

Figure 2.4
Intra-USCENTCOM Airlift Tons, by Mode, 2009

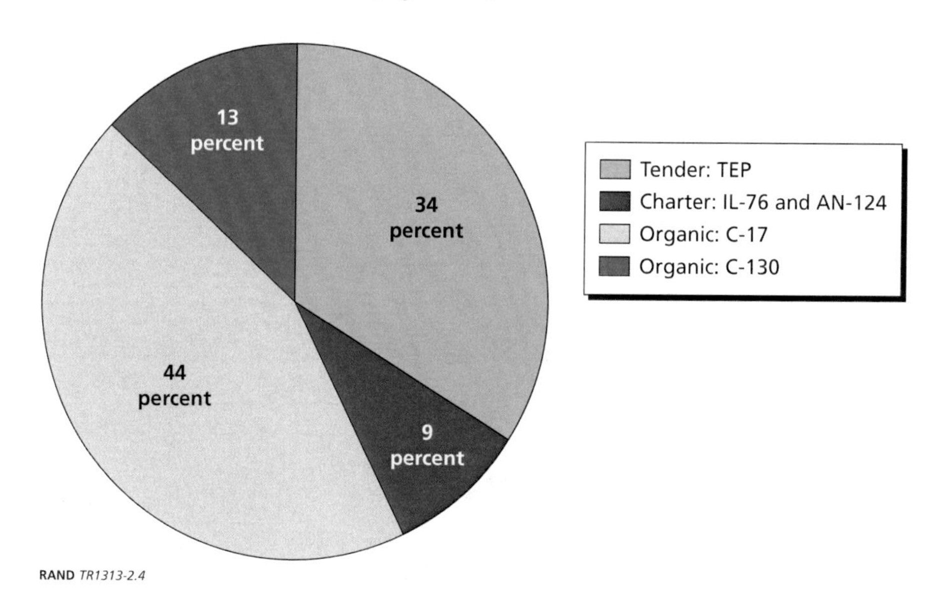

RAND *TR1313-2.4*

intra-USCENTCOM tonnage was transported via CITA, with TEP providing 34 percent of the movements and charters of IL-76s and AN-124s providing an additional 9 percent.[8] Figure 2.5 presents the temporal dimension of this same cargo tonnage data and demonstrates that, while TEP was utilized fairly extensively across the entire year, there was some variation

[8] Our data analysis identified 6,700 tons of intra-USCENTCOM cargo moved via other aircraft, primarily USAF C-5s, during 2009 (equal to 1.7 percent of the cargo total); these 6,700 tons do not appear in Figures 2.4 or 2.5.

Figure 2.5
Intra-USCENTCOM Weekly Airlift Tons, by Mode, 2009

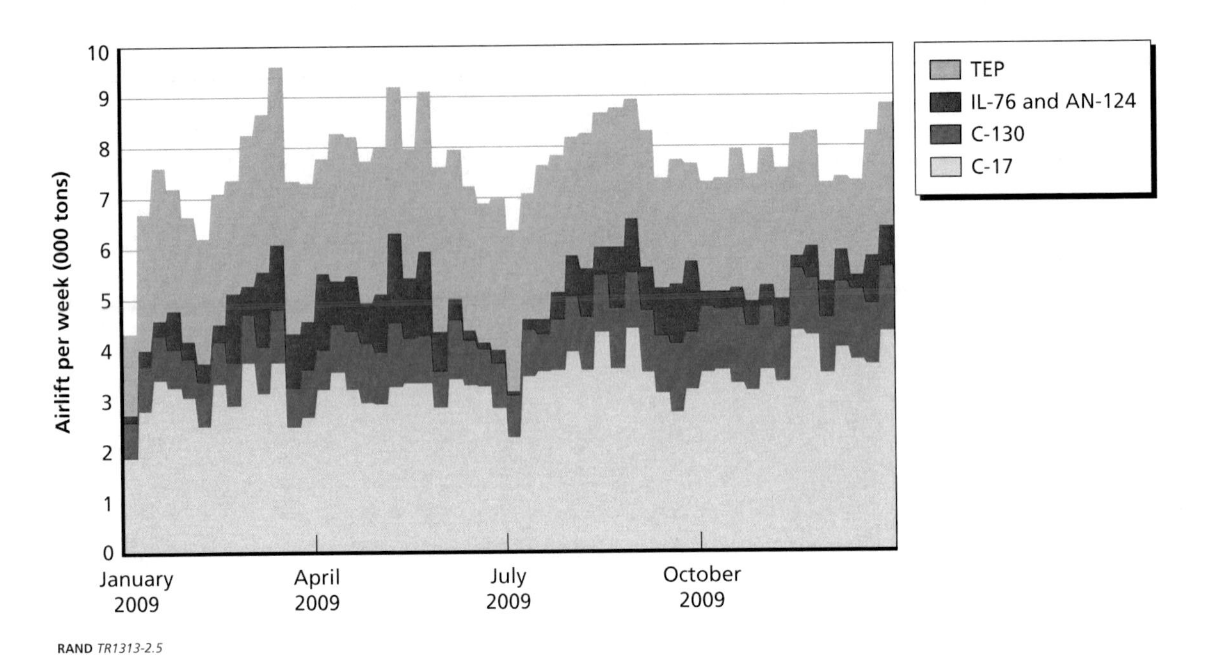

RAND *TR1313-2.5*

in the amount of charter movements, with peaks in April and October and troughs in July and January.

We also examined the characteristics of this set of cargoes. In addition to the motivations for using CITA that were discussed in Chapter One (extending the service life of the C-130 fleet, overcoming lack of access to ARC C-130 aircrews, reducing use of road convoys), both the IL-76 and AN-124 aircraft can transport large and heavy items that the C-130 cannot. The AN-124 is also able to accommodate some cargoes that cannot fit onto a C-17. Figure 2.6 is a scatter plot that presents the set of items transported intra-USCENTCOM via IL-76 and AN-124 charter in terms of distance transported (on the horizontal axis) and item weight (on the vertical axis). Each black dot on the plot represents an item moved via IL-76, and each red dot an item moved via AN-124. Note that any single dot may represent multiple items that have equal value along both dimensions. We have overlaid onto this plot the range-payload capabilities of the C-130 and C-17 (assuming no aerial refueling of the C-17),[9] with the region lying to the left of and below the dark blue line indicating the capability envelope of a C-130 and the light blue line similarly indicating the C-17's capability with no aerial refueling.[10]

Observe that many of the dots in Figure 2.6 appear to lie in vertical bands. This is because each vertical band here is associated with a specific distance, which generally corresponds to a specific origin-destination pair. Items that exceeded the C-130 capability curve were moved

[9] Lt Col David Cutter, Directorate of Analyses, Assessments and Lessons Learned, Headquarters Air Mobility Command, identified these range-payload capability curves using the Advanced Computer Flight Plan software. Note that these calculations assume no wind, a standard day, a step climb, and a requirement of one hour's worth of holding fuel at the destination.

[10] We present these range-payload curves to examine the size of individual items moved via charter versus C-130 and C-17 capabilities. The remainder of this analysis models airlift operations using *planning factors* to account for the frequency with which cargo consumes all available space on the aircraft before it reaches the maximum payload.

Figure 2.6
Items Moved Intra-USCENTCOM During 2009 via IL-76 and AN-124 Charters, Contrasted with C-130 and C-17 Range-Payload Capabilities

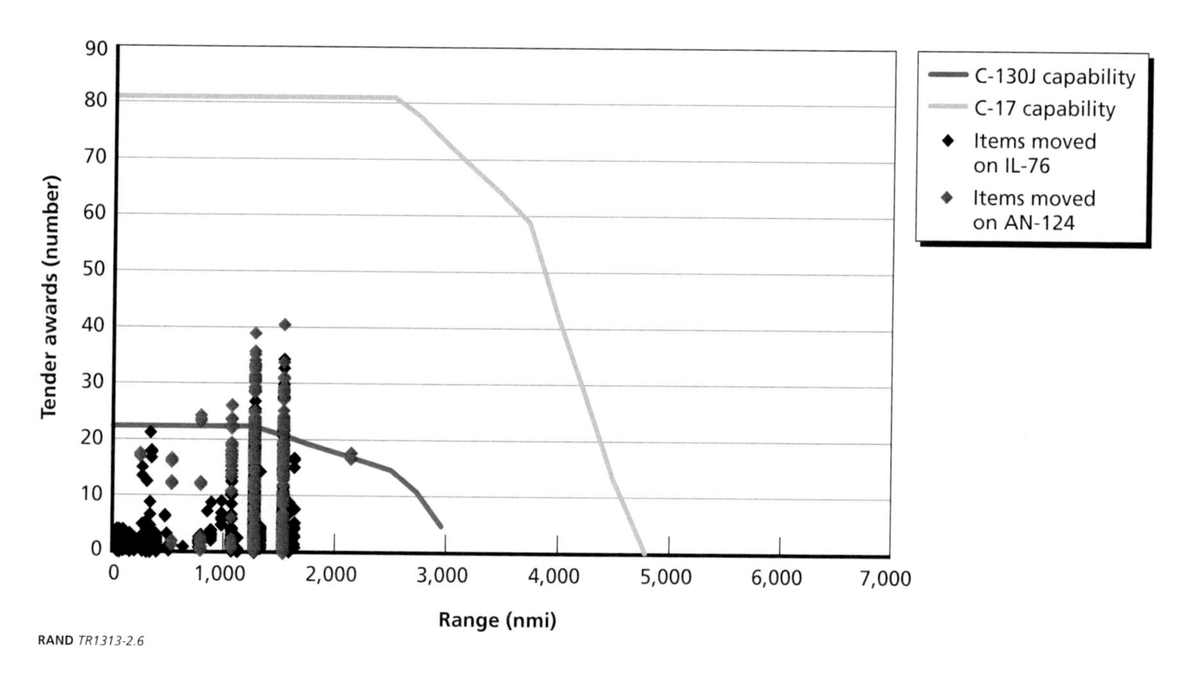

RAND *TR1313-2.6*

across five origin-destination pairs (in either direction, listed in order of increasing distance): Manas, Kyrgyzstan, to Kandahar, Afghanistan; Al Udeid, Qatar, to Kandahar, Afghanistan; Kuwait International to Kandahar, Afghanistan; Kuwait International to Bagram, Afghanistan; Kuwait International to Almaty, Kazakhstan. Only 5 percent of the total items in Figure 2.6 exceeded C-130 capabilities, and none came close to exceeding the range payload capabilities of C-17. Further, although an item can exceed an aircraft's volumetric constraints without violating the aircraft's weight limitations, all the items presented here consumed 7.3 pallet positions or less; thus, none exceeded the 18 pallet positions available on a C-17, and very few of them exceeded the six pallet positions of the C-130. The tallest item had a height of 170 inches, so none of these items would have exceeded the C-17's height restrictions. Therefore, in 2009, the IL-76 and AN-124 charters were not used to move intra-USCENTCOM items that exceeded C-17 capabilities.

We also examined the list of pallets and rolling stock transported via TEP and found that these items were relatively small, with 99 percent weighing 8 tons or less. As Figure 2.6 shows, 8 tons corresponds to a range of approximately 2,850 nmi on the C-130 curve, which is greater than the distance between Cairo, Egypt, and the easternmost point in Kazakhstan. Thus, TEP was also not used to move many items that exceeded the USAF-organic aircraft's capabilities.

In summary, the use of CITA in USCENTCOM in 2009 allowed USAF to maintain a relatively small mobility footprint of 13 C-17s and 29 C-130s deployed in theater, while reducing the amount of cargo requiring surface transportation. In the next chapter, we will examine whether this strategy was cost-effective.

CHAPTER THREE

Determining Cost-Effectiveness of CITA Movements

CITA cost-effectiveness can be evaluated by contrasting the cost to perform a set of ITA movements via commercial airlift with the cost to move the same cargo using USAF aircraft.[1] This requires the determination of appropriate "cost" metrics for each option. However, it is not readily apparent how one should construct such metrics, both in the selection of which data elements to include and in the form of the model that transforms these data elements into a cost.

For CITA, one can obtain the costs paid by the government to the commercial provider for a set of historical movements.[2] The data analysis we described in Chapter Two computed that TEP transported a total of 133,000 tons of cargo in 2009. We obtained the associated actual expenditures on TEP from the USTRANSCOM Acquisition Directorate; these costs totaled $382 million in 2009.

The metric *dollars per pound* is commonly used when discussing TEP as a measure of the cost-effectiveness of the program. From this perspective, TEP provided airlift movements at an average rate of $1.44 per pound. However, one criticism of this metric is that it does not account for the distance over which cargo is transported. Another commonly used metric for airlift cost-effectiveness is *dollars per ton-mile moved*. This set of TEP movements accounted for 103 million ton-miles,[3] equating to a cost of $3.69 per ton-mile moved.

One could compare this cost per ton-mile with what USAF-organic aircraft could achieve. Consider, for example, the fiscal year (FY) 2010 Transportation Working Capital Fund (TWCF) special assignment airlift mission (SAAM) rate of $12,317 per deployed C-17 flying hour (USTRANSCOM, 2010).[4] Assuming block flying speeds of 410 nmi per hour for the C-17 (Air Force Pamphlet [AFPAM] 10-1403, 2003), we obtained a ratio of $30.04 per

---

[1] Alternatively, if access to aircrews or aircraft were viewed as the most important constraint driving the use of CITA, cost-effectiveness could be evaluated in terms of the CITA cost per USAF aircraft not needed to be deployed. Our analytic formulation allows the examination of either approach. Note also the assumption that air transport was necessary for this set of cargoes and that surface transport was not an option; were this assumption relaxed, the ensuing multimodal analysis would require some basis for determining which cargoes "must" be airlifted and which could be moved by either mode.

[2] These *direct* expenditures do not account for all the government's costs associated with CITA; the establishment and oversight of the contractual relationships also levies a requirement on DoD personnel. Because we will not attempt to estimate such *transaction costs* in this analysis, our approach can be viewed as understating the total costs of commercial movements.

[3] We determined cargo ton-miles by computing the flying distance between each cargo's origin and destination, assuming that overflight of Iran was not permitted, measured in nautical miles.

[4] A midyear revision to SAAM rates reduced the C-17 cost to $10,280 per flying hour for missions occurring between July 1 and September 30, 2010. The letter announcing the rate reductions stated that rates would return to standard levels starting October 1, 2010.

mile. Using this logic, a C-17 sortie carrying 30 tons would do so at a rate of $1.00 per ton-mile.[5] A C-17 sortie payload of 8.14 tons would generate the same $3.69 per ton-mile rate that TEP obtained, on average; at higher payloads the C-17 would be more cost-effective, while at lower payloads the C-17 would be less cost-effective. Such a payload is comparable to the 8.66 ton average weight for a TEP tender during 2009; however, we observed that 15 percent of total tender awards shared a common origin, destination, and award date with another tender, rather than these multiple tenders being combined into a single large tender. As we will discuss later in this chapter, it is difficult to ascertain what C-17 payloads could have been achieved across this set of TEP movements without using an optimization model, because it is not apparent how these movements could have been connected into a series of C-17 sorties (including the potentially empty retrograde sorties that might be necessary following a movement). Nevertheless, it is within the capability of USAF-organic aircraft to perform at least some TEP movements at less cost than TEP.[6]

Moreover, determining how cost structures derived from these data would apply to a different set of movements requires careful analysis. In particular, it is necessary to identify the extent of *price elasticities of demand*: To what extent do changes in the demand for CITA movements affect the prices commercial carriers charge for these movements? This is the primary shortcoming with an attempt to draw conclusions about overall, *systemwide cost-effectiveness* by simply calculating the amount historically spent on each airlift option (even when normalized per ton-mile of cargo moved) and making a direct comparison across the alternatives. Such calculations can offer insight only into the relative cost-effectiveness of each alternative *subject to the set of cargo allocation decisions that were actually implemented*. Such an approach does not provide a means of determining whether a different decision that would change the allocation of some cargo would have reduced systemwide costs.

Furthermore, for USAF aircraft, the generation of cost metrics entails a different set of considerations. If the procurement (or retirement) of aircraft is under consideration, a set of *fixed costs*, such as maintenance manpower and aircrew training requirements, needs to be factored into the cost models. We, however, assumed that the decision to expand or reduce the use of CITA would not influence the USAF's fleet size requirement for C-17s and C-130s. Instead, we focused on the *marginal cost* associated with varying the number of FH performed by aircraft that are already in the USAF inventory.

A Cost Model for TEP Movements

We first turn to the development of a cost model for TEP tender movements. Because TEP does not operate under predetermined costs to move cargo between specific origin-destination pairs, it was necessary to develop a model that could generate TEP costs for any potential ITA movement.

[5] In Figure 2.6, 30 tons is significantly less than the maximum C-17 payload for a sortie of 3,500 miles, which is farther than the distance across the USCENTCOM AOR.

[6] For example, consider three TEP tenders that were awarded on June 2, 2009: (a) Kuwait International to Bagram, 21.64 tons, $43,272; (b) Kuwait International to Bagram, 16.47 tons, $32,948; and (c) Bagram to Kuwait International, 17.91 tons, $35,816. The total cost for these tenders is $112,036. Assuming overflight of Iran is not permitted; the distance between these sites is 1,533 nmi. Assuming one C-17 sortie in each direction, at 410 nmi per hour, these movements would require 7.48 C-17 flying hours (FH) at a TWCF SAAM cost of $92,131.

The USTRANSCOM Acquisition Directorate provided us with a TEP data set covering the interval from October 1, 2008, to December 17, 2009. This data set contained information on all tenders that were offered to commercial carriers; the bids that were received from each of the seven carriers that participated in TEP at that time, along with an identification of the winning bid; and pallet-level details on the movement history of all items moved through TEP. Across this entire data set, over 16,000 tenders were awarded, at a total cost of $400 million; over 88,000 pallets and 154,000 tons were moved on this set of tenders. These tender movements took place over a set of 284 routes, covering 34 locations across 11 countries. All locations TEP served were within the USCENTCOM AOR, with one exception, Djibouti-Ambouli International Airport, which is in the USAFRICOM AOR.

We observed significant variation across the tender bids and across the award amounts, even for similar movements across a common route. Consider the example in Table 3.1, which compares three tenders that each moved between Kuwait City and Djibouti. All three movements occurred within four days of one another in May 2009, and all were comparably sized, ranging between 19 and 27 tons. The bids that were received for these movements varied significantly, even for a single carrier. For example, carrier A offered a bid of $177,000 on the first movement; $89,000 on the next movement; and $77,000 on the third movement. Moreover, for the May 16, 2009, movement, carrier A's bid of $77,000 was selected, even though it was the second highest of the bids received. Across 2009, the lowest bid was selected approximately 60 percent of the time. As we discussed in Chapter Two, bids are evaluated not only on the basis of cost but on a combination of cost and some consideration of the carrier's past performance.

We conducted a regression analysis to identify the factors that influenced the cost of awarded TEP tenders in 2009, examining such factors as the number of bidders, the cargo

Table 3.1
TEP Example of Three Similar Tender Movements Between Kuwait City and Djibouti Ambouli

	Tender 1	Tender 2	Tender 3
Tender date	May 13, 2009	May 13, 2009	May 16, 2009
Tons to move	26.6	26.0	19.0
		Bids ($)	
Carrier A	177,354	89,406	76,561
Carrier B	119,475	93,564	85,703
Carrier C	99,828	100,321	75,037
Carrier D	NA	NA	NA
Carrier E	88,146	88,366	62,849
Carrier F	81,774	80,049	65,896
Carrier G	NA	NA	69,705
Carrier selected	F	F	A

weight, flight time, and the cargo origin and destination pairs. The following were the primary findings from the regression analysis:

- *Competition is important for reducing tender cost.* A 10-percent increase in the number of bidders on a tender causes tender costs to decline between 3.8 and 5.5 percent.
- *Carriers prefer to bid on more active routes.* Routes with 10 percent more tenders had, on average, between 1.4 and 1.9 percent more carriers bidding on cargo movements.
- *Cargo bundling can reduce costs.* As the weight of a given tender increases 10 percent, the costs per pound transported decline by between 2.7 and 3.0 percent.
- *Differences in flight time explain some of the cost variation between routes.* As flight times increase 10 percent, tender costs increase by between 3.0 and 4.2 percent.
- *The specific location of the cargo origin and destination are also important predictors of tender cost.* The influence extends beyond just the flight time between the origin and destination pair.

We used the statistical relationship between cost and origin-destination pair serviced, tender size (measured by the weight of the cargo tendered), and the number of past tenders on the route to estimate TEP costs. We assumed that these factors interact multiplicatively to determine tender costs.[7]

We also examined the timeliness of delivery of TEP movements. Figure 3.1 presents the probability distribution of TEP tenders by the delivery time achieved. The green and blue bars indicate differences in the tender delivery time requirements. Initially, tenders had a 72-hour delivery requirement, but between June and November 2009, two tender priorities were used:

Figure 3.1
TEP Delivery Performance

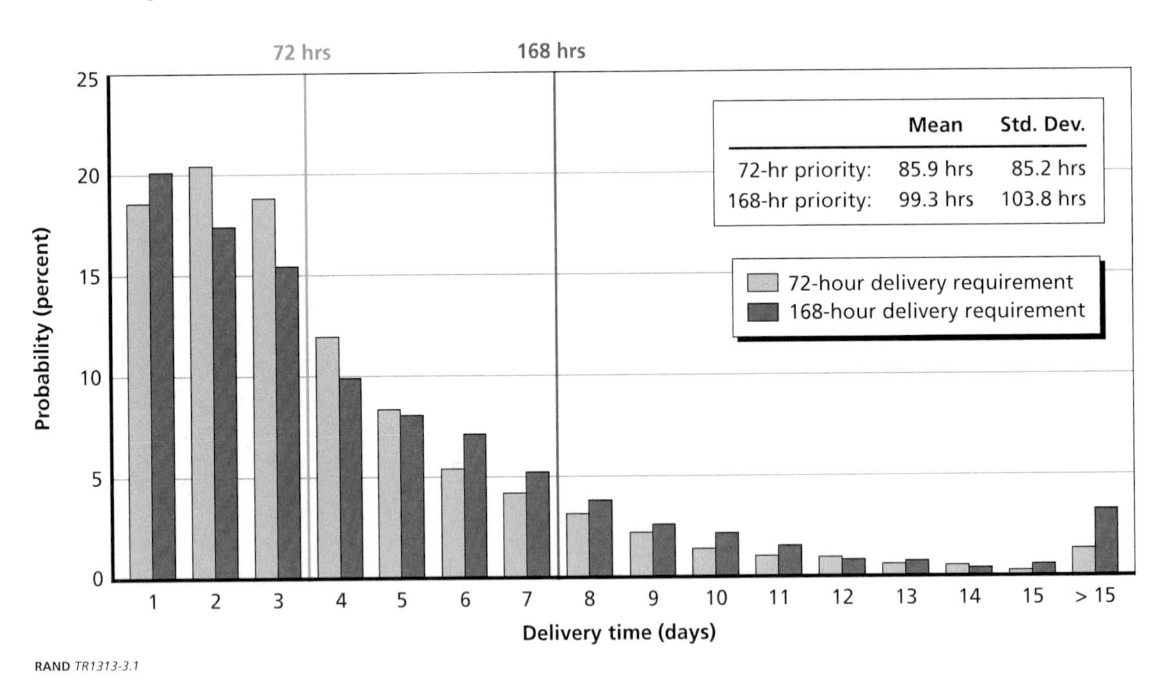

RAND *TR1313-3.1*

---

[7] See Appendix B for the formal details of the regression analysis and the interpretation of the coefficients.

A more urgent 72-hour requirement was applied to 85 percent of the tenders, with a less-urgent 168-hour requirement applied to the remaining tenders. After November 2009, the different priorities were eliminated, and all tenders again faced a 72-hour delivery time requirement.

Focusing on the leftmost green bar, approximately 19 percent of all deliveries with a 72-hour delivery requirement were delivered within one day. Summing across days one through three, approximately 60 percent of the deliveries with a 72-hour delivery requirement satisfied that requirement. Of the 40 percent that were delivered later than the requirement, 21 percent were granted "exemptions" for a variety of reasons, including weather, runway closures, airport construction, or loading and unloading issues.

The general shape of the distribution represented by the green and blue bars is very similar. Regression analysis conducted to isolate the effect on cost of varying the delivery time requirement suggests that, as opposed to the 72-hour requirement, the 168-hour delivery time requirement did not reduce tender costs in any statistically measurable way and delayed average delivery times by less than 24 hours. These findings support the decision to eliminate the two delivery time requirements and return to a single 72-hour requirement, since the use of a longer requirement did not achieve a cost reduction.

Cost of Charter Movements

Rather than develop a cost model for aircraft charters in this analysis, we instead applied the contract language appearing in USTRANSCOM's FY 2010 IL-76 contract with Silk Way Airlines to all potential ITA charters. This contract gave specific prices for IL-76 movements between certain origin-destination pairs (e.g., $134,000 for a sortie between Kuwait International and Bagram, Afghanistan); an $89-per-mile rate for movements between other locations within USCENTCOM; and a $17,000 cancellation charge per mission not utilized below the contract's guaranteed level (which, for this contract, was defined as two daily missions between Kuwait and Afghanistan, four daily missions between Kuwait and Iraq, or one daily Kuwait-Afghanistan mission and two daily Kuwait-Iraq missions).

Cost Model for USAF Aircraft

The fairest possible comparison between the commercial cost to move ITA cargo and the cost to move the same cargo on USAF aircraft needs to consider the total marginal cost of performing the movement via a C-130 or C-17. This total cost should include direct costs, such as fuel and consumables, as well as the costs (and potential benefits) associated with speeding up major maintenance and aircraft replacement decisions.

We generated such marginal cost estimates for C-130 FH by using a discounted cash flow model.[8] This model computes the present discounted cost of maintaining C-130 capabilities indefinitely, with aircraft being maintained and replaced over time to maintain a constant fleet size. This analysis looks at the fleet of C-130E and C-130H aircraft and assumes that, in the future, they will be replaced by C-130Js. The model was based on a June 2009 snapshot of the C-130E and C-130H fleet and uses data on each aircraft's equivalent baseline hours (EBH, a

---

[8] Appendix C details this analysis.

metric that tracks an aircraft's remaining operating life), historical home-station and deployed flying activity (including for each both the hours flown and the associated severity factor [SF]), and projected future aircraft utilization.[9]

We identified operating costs based on the Cost Oriented Resource Estimating (CORE) model, which uses standard USAF planning factors, published in appendixes to Air Force Instruction (AFI) 65-503, to estimate the marginal cost of adding a squadron to (or removing one from) an existing active-duty base. Cost categories in the CORE model are classified as either variable or fixed. *Variable costs* are those associated with additional flying (i.e., those that would change if a cargo mission were added to a deployed force), such as fuel consumption, consumables, and some engine overhaul costs. The variable costs differed significantly for each mission design series (MDS), ranging from $5,900 per C-130H FH to $3,300 per C-130J FH. C-130Js have a significantly lower variable cost than do the older models. To the extent that additional deployed flying accelerates replacement dates for older aircraft, increasing their deployed FH allows the USAF to enjoy the lower costs associated with C-130J at an earlier point in time.

The fixed costs in the CORE model account for the costs of maintaining the aircraft at either home station or a deployed location. When an ARC aircraft is deployed, some deployed personnel must move to activated status, which increases fixed costs. These fixed costs varied by major command (MAJCOM) and MDS. The C-130J also has lower fixed costs per aircraft than does the C-130E or C-130H.

In addition to the operating costs obtained from CORE, we included costs associated with major maintenance activities that are triggered at certain EBH points for each aircraft, namely rainbow fittings maintenance costing $700,000 at 24,000 EBH and Time Compliance Technical Order (TCTO) 1908 center-wing inspection at 38,000 EBH (Orletsky et al., 2011). Finally, aircraft replacement requires procuring new C-130J aircraft; we assumed that this occurs for any aircraft at 45,000 EBH and costs $63.9 million (based on C-130J flyaway cost estimates from Orletsky et al., 2011).

We estimated the marginal cost for additional deployed C-130 FH, taking into account how the additional flying affected the timing of major maintenance costs, aircraft replacement costs, and the differential fixed and operating costs of the C-130E and C-130H versus their replacement—the C-130J. Thus, to the extent that an additional FH hastens the replacement of an older aircraft with a C-130J, we identified the change in total present value that is due to the change of replacement date. Time-discounted aircraft procurement and major maintenance costs would increase as a result, but future fixed and variable costs would decrease because the USAF can enjoy the C-130J's operating cost advantages earlier.

When computing these discounted cash flows, we discounted future costs to FY 2010 dollars using the long-term real discount rate of 2.7 percent per year prescribed by the Office of Management and Budget (OMB, 2009) and assumed no real cost inflation.

Table 3.2 presents our marginal cost estimates for an additional deployed flying hour, computed in FY 2010 dollars and differentiated by MAJCOM (based on differences in the composition of each MAJCOM's fleet of C-130E and C-130H aircraft and on differences in deployment costs for active-duty and ARC units). An additional deployed C-130 flying hour

[9] We based these cost calculations on the C-130E and C-130H fleet because of the lack of data for the newer-model C-130J. Because C-130Js are less expensive to operate per hour than the older C-130E and C-130H, our cost estimates can be viewed as somewhat higher than USAF fleetwide average cost to perform a deployed C-130 flying hour.

Table 3.2
Marginal Cost Estimates for a One-Time Deployed Flying Hour

Command	Cost ($ FY 2010)
Air Force Reserve Command (AFRC)	6,600
Air Mobility Command (AMC)	5,800
Air National Guard (ANG)	7,300
Pacific Air Forces (PACAF)	7,000
U.S. Air Forces in Europe (USAFE)	6,900
Overall average	6,800

costs, on average, $6,800 on a discounted cash flow basis. Of this total, $5,900 was associated with one-time direct costs (the variable costs in the CORE model); the remaining $900 was associated with the present discount value of speeding up major maintenance activities and aircraft replacement, offset by any savings that may occur due to accelerating the date at which the operating costs for older C-130 variants are replaced with those for the lower-operating-cost C-130J.

As a point of comparison, our estimated cost of $6,800 per FH is very similar to the $6,967 per C-130E/H FH charged to DoD users via the FY 2010 TWCF SAAM rate (USTRANSCOM, 2010).[10] We did not generate marginal cost estimates for C-17 FH because of uncertainties about future major maintenance actions for this aircraft and a lack of information about its fleet age. Instead, because the TWCF DoD SAAM rate was so similar to our discounted cash flow calculations for the C-130, we directly used the FY 2010 TWCF DoD SAAM rate of $12,317 per deployed C-17 FH.

Optimization Model

Thus far, this chapter has described our development of cost models for the use of CITA and USAF aircraft. However, to make cost-effectiveness determinations, it is also necessary to identify how ITA movements translate into requirements for aircraft sorties for IL-76 charters,[11] C-130s, and C-17s. Such a translation into sorties is not necessary for TEP movements because our TEP model simply allows cost to vary as a function of the origin-destination pair, the amount of cargo moved on a tender, and the number of tenders made on the route in the past.

As a cost-effectiveness check, the TEP uses TWCF FH rates to estimate the cost for a C-17 or C-130 to perform each tendered movement, assuming that each tendered movement

[10] A midyear revision to SAAM rates reduced the C-130E/H cost to $5,815 per FH for missions occurring between July 1 and September 30, 2010. The letter announcing the rate reductions stated that rates would return to standard levels starting October 1, 2010.

[11] Our optimization model does not include the chartering of AN-124 as a delivery option but instead assumes that all charter movements would be performed on an IL-76. This reduced the dimensionality of the optimization model but did not introduce limitations on the feasibility of airlift as a delivery option for any cargo because, as discussed in Chapter Two, every item moved via AN-124 within USCENTCOM during calendar year 2009 could have been flown on a C-17.

would require a separate sortie. Using this process, TEP estimates that it would have cost $610 million to perform the set of 2009 movements via the less expensive of the C-17 or C-130, based on the characteristics of the movement. That it cost $380 million for TEP to provide these movements suggests that TEP achieved a cost avoidance of $230 million. However, the assumption that each tender movement would require a separate sortie may not be reasonable. Most of these tenders were for relatively small movements, with 48 percent for movements of 5 tons or less and 79 percent for movements of 15 tons or less. Note that such weights are significantly less than the capabilities of a C-17 or C-130.[12] Moreover, as discussed previously, it may be possible to combine multiple tender movements along a common origin-destination pair because 15 percent of total tender awards shared a common origin, destination, and award date with another movement (similar to tenders 1 and 2 in Table 3.1). Alternatively, it might be possible to string multiple origin-destination pairs together into a single mission, based on such factors as the required delivery dates and the distances and flight times between specific locations. This suggests that TEP's cost avoidance values are likely an inaccurate portrayal of the program's performance.

Making the most accurate comparisons between the cost to perform a set of ITA movements via commercial airlift and the cost to move the same cargo using USAF aircraft requires solving an integrated routing and assignment problem: Which aircraft should fly which missions, and how should movements be assigned to those missions? We developed an optimization model to solve this problem.

Our optimization model is a large-scale mixed integer linear program (MILP) that uses integer assignments for such factors as the assignment of aircraft to routes and the choice of whether to tender via a TEP movement. The model's objective is to minimize the total cost of airlift over the modeled time horizon, including the operating costs for military assets, funds spent to acquire leased aircraft, and the cost to purchase TEP movements. The model was coded using the General Algebraic Modeling System, running the commercial optimization solver Cplex.[13] This optimization model was based on the framework that was developed in the Naval Postgraduate School–RAND Mobility Optimizer (Baker et al., 2002).

Appendix D details the model's formulation. There are two significant differences between the modeling formulation we employed here and the optimization models typically used for such problems. First, based on the findings of our regression analyses of TEP costs, our model allows TEP costs to vary *dynamically* according to three parameters: the distance between the cargo's origin and its destination, the cargo's weight, and the frequency of past tender usage across the origin-destination pair. Second, our model does not optimize across the entire time frame under consideration at one time, which is unrealistic given the limited horizon over which planners have knowledge of future cargo requirements. Instead, the model determines the most cost-effective allocation of cargo to aircraft and aircraft to routes within a two-day window. Given a solution for the two-day window, t_1 to t_2, the model then executes the movements scheduled for day t_1, records the optimal movements for day t_2, and advances the time

[12] AMC planning factors published in AFPAM 10-1403 state that the allowable cabin load for a C-17 is 65 tons (assuming a 3,200-nmi flight leg) and that the load for a C-130 is 17 tons (assuming a 2,000-nmi flight leg).

[13] All model runs were performed on an eight-core workstation (to capitalize on Cplex's ability to multithread a computer's individual cores and expedite solution times). In the most computationally challenging scenarios we examined, the model required approximately one week of clock time (or 50 days of central processing unit [CPU] time) to run to completion.

horizon by one day. The MILP now has access to a near-feasible starting point for period t_2 and uses it to initiate the next model run, which now extends from t_2 to t_3.

Thus, for a fixed number of deployed USAF aircraft, the model identifies an optimal utilization of the aircraft that achieves the minimum-attainable total ITA delivery cost. Parametrically varying the deployed USAF fleet sizes reveals how additional or fewer C-17s or C-130s affect cost. This gives an alternative perspective on CITA cost-effectiveness to be evaluated using our optimization model: the cost spent on CITA per USAF aircraft not needing to be deployed. Such a metric could address aspects of the second motivation offered in Chapter One for the use of CITA (lack of access to ARC aircrews or aircraft) by identifying the number of additional USAF aircraft that would have been needed in USCENTCOM to execute the CITA movements.

CHAPTER FOUR

Results

We applied the optimization models described in Chapter Three to the set of 2009 USCENTCOM ITA cargo and passenger movements to determine whether the expenditures on commercial movements were cost-effective, relative to the cost of having USAF-organic aircraft perform these movement.[1] In the model, we allowed for four different airlift options: C-130s, C-17s, TEP tenders, and IL-76 charters.

Table 4.1 presents the total amount of intra-USCENTCOM cargo and passengers that each of these airlift options moved in 2009, computed as described in Chapter Two. This table also presents the ton-miles and passenger-miles associated with each airlift option, which we determined by computing the flying distance in nautical miles between the origin and destination of either each tender (for TEP) or each sortie (for all other airlift options), assuming that overflight of Iran was not permitted.[2] The 38,000 tons and 52,000,000 ton-miles of IL-76 cargo presented here actually represent the total ITA cargo that both IL-76 and AN-124 charters carried. Other aircraft, primarily USAF C-5s, moved another 6,700 tons and 7 million ton-miles of intra-USCENTCOM cargo in 2009 (equal to 1.7 percent of the total cargo tonnage, 2.0 percent of the total cargo ton-miles). While Table 4.1 does not include the data for these other aircraft, the ITA movement requirements inputs to the optimization model did include this cargo. Thus, the movement totals associated with each optimization model solution include this additional 6,700 tons of cargo.

Table 4.1
Total Intra-USCENTCOM ITA Movements in 2009, by Airlift Type

Airlift Type	Cargo Tons (000s)	Passengers (000s)	Cargo Ton-Miles (000s)	Passenger-Miles (M)
C-17	173	542	167	346
C-130	54	592	18	219
TEP	133	—	103	—
IL-76	38	—	52	—

[1] Because our TEP data set did not include movements for the last two weeks of 2009, we based our data set's tender movements for December 18–31, 2009, on the TEP movements performed during December 18–31, 2008 (for which we had data).

[2] The ton-miles associated with TEP thus reflect a *requirement*, while the ton-miles associated with all other airlift options reflect the *set of actual sorties performed to satisfy a requirement*.

We based the cost structure for each airlift option on the analyses in Chapter Three. USAF-organic aircraft were assumed to cost $6,800 per C-130 FH and $12,300 per C-17 FH. We used the regression-based estimates of TEP tender costs, with the cost varying depending on the origin-destination pair; the amount of prior TEP activity over that origin-destination pair; and the size of the shipment, with one cost for shipments between 0 and 5 tons, another cost for shipments between 5 and 15 tons, and another cost for shipments between 15 and 25 tons. The IL-76 charter costs were based on the FY 2010 contract with Silk Way Airlines.

While the number of C-130s and C-17s available for ITA movements is fixed at a user-input level, the number of chartered IL-76 is not similarly constrained. Instead, the model can elect to charter as many IL-76 as it deems necessary on any given day, but it will then be assumed that this many aircraft are on contract and must be utilized on all subsequent days. If fewer IL-76s are tasked on a later date, the $17,000 cancellation charge per aircraft not utilized is applied. As discussed previously, we did not include a cost associated with the deployment of C-17s or C-130s to the theater (but we will discuss these costs briefly later in this chapter).

We assumed that all passenger movements must occur on either a C-130 or a C-17. We assumed that each C-17 sortie had a maximum payload of 45 tons or 188 passengers or of some linear combination of cargo and passengers between these two endpoints. We assumed that each C-130 sortie had a maximum payload of 12 tons or 92 passengers or of some linear combination of cargo and passengers between these two endpoints.[3] We assumed that each IL-76 had a maximum capacity of 44 tons, with no passenger capacity. We assumed block flying speeds of 410 nmi per hour for both the C-17 and IL-76 and of 270 nmi per hour for the C-130.[4] Because of the nature of TEP tenders, for which DoD is contracting only on the movement of an item and is not monitoring *how* the item gets from its origin to its destination, we did not build a fleet of TEP aircraft into the model or track each TEP aircraft's movement through the system. Instead, we assumed that each TEP movement would arrive at its aerial port of debarkation one day after being tendered and picked up by the delivery provider.

The optimization model requires as an input a set of allowable aircraft routes for the C-130, C-17, and IL-76. We assumed that all C-130, C-17, and IL-76 aircraft must begin and end their flying day at one of the following eight locations: Bagram, Afghanistan; Kandahar, Afghanistan; Al Sahra, Iraq; Balad, Iraq; Ali Al Salem, Kuwait; Manas, Kyrgyzstan; Thumrait, Oman; and Al Udeid, Qatar. An aircraft could start and end its day at two different locations. In a practical sense, this is a set of midsize to large regional bases that possess sufficient security and maintenance to protect and service a significant subset of the mobility fleet. We limited the set of allowable sortie origin-destination pairs for all three aircraft types to those that a C-130 or C-17 actually flew in 2009. We generated missions from this set of allowable sorties by stringing together sets of up to three sorties, provided that the total mission time (including quick-turn ground times) did not exceed a 16-hour duty day. Thus, even if we could identify four relatively short sorties that would fit within a mission duty day constraint, we did not allow the model to conduct such a mission. For all aircraft, any positioning or depositioning flights needed would have to fit within this route structure, and such movements incurred costs at the same rate as "live" cargo-hauling movements and consumed part of the aircraft's allow-

---

[3] The maximum cargo capacities for the C-17 and C-130 correspond to the cargo planning payloads in AFPAM 10-1403. The passenger capacities assumed palletized seating.

[4] The block speeds for USAF-organic aircraft came from AFPAM 10-1403. The IL-76's block speed is from Skyline Aviation Ltd., "Ilyushin IL-76," online, undated.

able duty day. TEP movements were not similarly constrained by a route structure and were simply allowed to occur between any origin-destination pair within USCENTCOM.

To determine the costs associated with the historical movements in Table 4.1, we ran this set of ITA movements through our optimization model, forcing all cargoes to move on the airlift type actually used in 2009.[5] Table 4.2 presents the optimization model's estimates of the costs associated with each airlift option. The table also contrasts these cost estimates with the actual expenditures on TEP and IL-76 charters, obtained from data provided by the USTRANSCOM Acquisition Directorate.[6] For both CITA options, our model overestimated the cost of the actual movements, by 20 percent for TEP and by 5 percent for aircraft charters. The difference associated with TEP movements is primarily due to our model's discretized cost structure, which assumes, for example, that all tender movements between 0 and 5 tons incur the cost of a 5-ton shipment. By multiplying the USAF-organic cost per FH values computed previously by the actual FH reported in GDSS for this set of missions, we obtained a comparable "actual 2009 expenditures" value for the C-17 and C-130.[7] For USAF-organic aircraft, our model underestimated the total cost of the actual movements by 10 percent. This difference was primarily due to the fact that the actual FH, as reported in GDSS, frequently are larger than the FH that the aircraft block speeds used in our model imply.[8]

Optimization Using Assets Available in USCENTCOM in 2009

We first determined the minimum level of C-17s and C-130s necessary to support the USAF-organic-only passenger movements. Setting the number of employable C-17s equal to 15, the 2009 daily average in USCENTCOM, we found that it took a minimum of 21 C-130s to support the peak day passenger movement requirements. Running the model with 15 C-17s and 20 C-130s returned an infeasible solution because the model was unable to satisfy the set of passenger movement requirements. Note that this minimum necessary number of C-130s, given 15 C-17s, happened to be equal to the average number of C-130s that were performing intra-USCENTCOM sorties across each day in 2009.

Table 4.2
Total Cost for Historical Intra-USCENTCOM ITA Movements in 2009, by Airlift Type

Cost Type	Total Cost ($M)	C-17 ($M)	C-130 ($M)	IL-76 ($M)	TEP ($M)
Optimization model estimate of 2009 experience	1,234	348	202	224	460
Actual 2009 expenditures	1,209	378	234	214	382

[5] For this model run, we forced all cargo that actually moved via C-5s onto C-17s and all cargo that actually moved via AN-124s onto IL-76s.

[6] The $214 million for IL-76 Actual 2009 expenditures includes costs associated with historical use of AN-124 charters.

[7] The FH associated with the 6,300 tons transported intra-USCENTCOM by C-5s were included in the C-17 costs and were assumed to incur costs at the C-5 TWCF DoD SAAM rate of $26,988 per flying hour; these C-5 FH costs accounted for a total of $16 million.

[8] Later in this chapter, we discuss our sensitivity analyses, which scaled the TEP cost structure by 4/5 and the other modes' cost structures by 5/4 to see how these variations affected our findings.

However, running the optimization model with 21 C-130s and 15 C-17s produced a peculiar result in the model solution. While the objective function value was considerably less than the $1.2 billion of actual 2009 expenditures, there was an exceedingly large expense associated with IL-76 cancellation fees. On the peak day's demand (which occurred on day 225), 42 IL-76s were chartered in this model run. Because the model had now placed 42 aircraft on contract, it was required either to use this many IL-76s on each subsequent day or to pay a cancellation fee. However, because the demand was much lower than this peak level on almost all subsequent days, an average of over 27 chartered IL-76 aircraft were idle from day 226 onward. This result was due to the model's cost structure, which attempts to minimize costs over a two-day window (the current day, plus the demands for delivery by tomorrow that are currently known). In the event of a peak day's demand, the model has a myopic perspective that attempts to satisfy this peak demand at minimum cost. Such a perspective can lead to the chartering of a large number of aircraft that may be underutilized in the future. The model structure has no incentive to avoid the potential for future penalties associated with underutilized charters by paying more in the near term to move this atypically large demand via a tender program. Were this model to be developed further, it could be improved by adding an ability to gauge the extent to which a day's demands are "unusually" large and to consider the long-term effects of entering into less-flexible chartering arrangements, even if such chartering arrangements would minimize costs over the near term, all the while recognizing the uncertainty associated with future levels of demand.

Because this solution generated such seemingly large IL-76 cancellation costs, we examined an alternative model, in which we imposed an upper bound of 24 IL-76 charters. In this model's solution, the days that had previously utilized more than 24 charters (there were five such days) instead used TEP to move the excess cargo that could no longer be shipped via IL-76. Because the entire fleet of C-130s and C-17s was already in use on these peak days, the USAF-organic aircraft were not available to provide these movements. When we compared the objective function values, the "unconstrained IL-76" solution cost $1,109 million, while the "upper bound on IL-76 charters" solution cost $1,034 million. The imposition of an upper bound of 24 chartered aircraft reduced the objective function value by $75 million, primarily by reducing IL-76 cancellation fees.[9]

Figure 4.1 presents the total delivery cost for both the historical experience and the "upper bound on IL-76 charters" optimization model's solution, allowing the model to employ 15 C-17s and 21 C-130s per day, which is equal to the average number of these aircraft that were performing ITA missions per day in USCENTCOM in 2009. Table 4.3 presents further details on the cost, cargo, and passengers each airlift option moves under each solution.[10]

Comparing the actual experience, with a total delivery cost of $1,209 million, against the best possible performance that can be achieved using a similar number of C-17s and C-130s reveals that the optimization model was able to reduce the total delivery cost by $175 million.

---

[9] We did not enforce an upper bound on the number of IL-76 charters on any of the other optimization model runs discussed in this report because we did not observe these peculiarities in any other solution.

[10] As discussed previously, the optimization solution column shows a total of 404,000 tons being transported, as opposed to the 398,000 tons in the 2009 experience column, because the optimization solution required the combination of these four airlift types to transport the 6,300 ITA tons that C-5s moved in the historical data. These 6,300 tons accounted for 7 million ton-miles of airlift movements in the 2009 experience. As in Table 4.2, the $16 million cost of these C-5 movements is included here in the 2009 experience column C-17 cost.

Figure 4.1
Total Delivery Cost for a Level of USAF Resources Equal to Actual USCENTCOM Usage in 2009

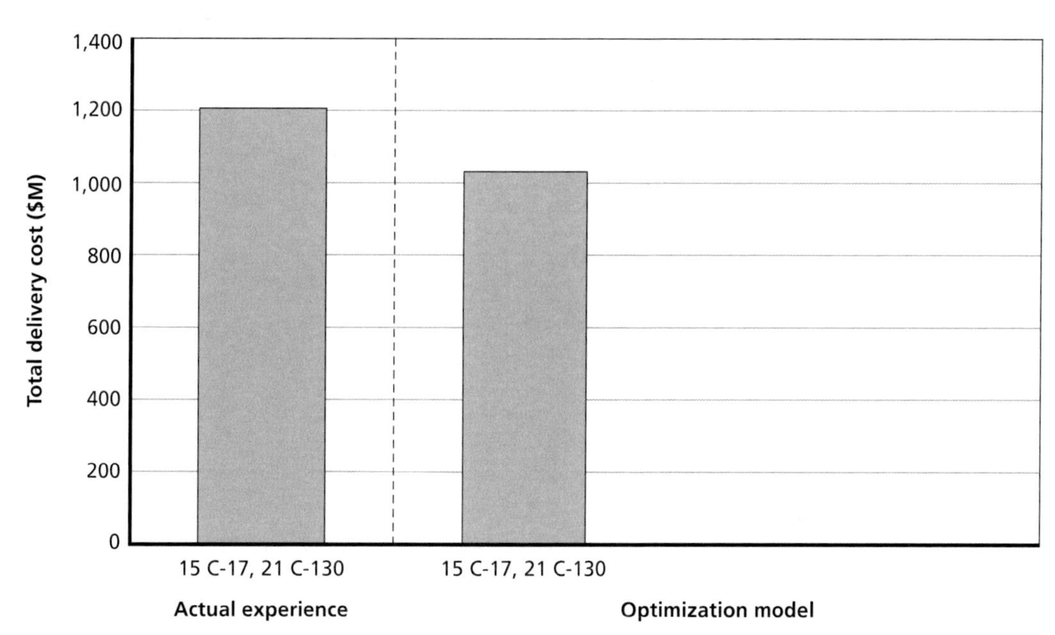

RAND *TR1313-4.1*

In the aggregate, the model solution slightly modified the total amount of cargo moved on C-17s and C-130s and significantly decreased cargo movement on TEP while greatly increasing the use of IL-76 charters. However, an examination of the movements at the level of origin and destination countries reveals clearer patterns that explain how the model was able to generate such large savings.

Table 4.4 presents the difference between the optimization model solution and the historical experience, differentiated by airlift type, for three origin-destination country pairs, computed as the total tonnage transported in the optimization model's solution minus the total tonnage actually transported in 2009. In this table, a positive value thus indicates that the optimization model increased the use of a particular airlift type across an origin-destination country pair, and a negative value indicates that the optimization model reduced the amount of cargo transported by that airlift type over the country pair. The optimization model solution made two large-scale substitutions, greatly decreasing TEP and increasing C-17 use across long movements, such as between Kuwait and Afghanistan (TEP decreased by 15,000 tons, and C-17s increased by 20,000 tons), and greatly decreasing TEP and increasing IL-76 use across short movements, such as intra-Iraq movements (TEP decreased by 20,000 tons, and IL-76s increased by 26,000 tons) or movements between Kuwait and Iraq (TEP decreased by 25,000 tons, and IL-76s increased by 18,000 tons).

As an illustration of the model's motivation for replacing TEP with C-17s for long movements, consider the following route: Bagram, Afghanistan, to Kuwait International to Camp Bastion, Afghanistan, to Bagram. Given the distances between these locations and with no overflight of Iranian territory, we estimate that it would cost approximately $94,000 to fly a C-17 mission across this route. And, in fact, the optimization model chose to perform 78 such missions, each carrying on average 7.5 tons between Bagram and Kuwait International, 5.9 tons between Bagram and Camp Bastion, 6.0 tons between Kuwait International and

Table 4.3
Optimization Model Solution Versus Historical Experience, Each Utilizing an Equal Number of C-17s and C-130s

Airlift Type	2009 Experience					Optimization Solution				
	Cargo (000 tons)	Passengers (000s)	Cargo Ton-Miles (M)	Passenger-Miles (M)	Cost ($M)	Cargo (000 tons)	Passengers (000s)	Cargo Ton-Miles (M)	Passenger-Miles (M)	Cost ($M)
C-17	173	542	167	346	378	183	621	207	411	303
C-130	54	592	18	219	234	43	512	19	209	157
TEP	133	—	103	—	382	41	—	43	—	182
IL-76	38	—	52	—	214	137	—	85	—	392

Table 4.4
Optimization Model Solution Versus Historical Experience, by Airlift Type and Origin-Destination Country Pairs

Airlift Type	Optimization Model Minus 2009 Experience, by Origin-Destination Country Pairs (000 tons of cargo)		
	Iraq–Iraq	Kuwait–Iraq	Kuwait–Afghanistan
C-17	–12	+6	+20
C-130	+6	0	0
TEP	–20	–25	–15
IL-76	+26	+18	–5

Camp Bastion, and 36.2 tons between Kuwait International and Bagram. In addition, the model also elected to move 1,200 total passengers across these 78 missions. In contrast, our model estimates the total TEP cost would be approximately $107,000 to $123,000 to move slightly less cargo and no passengers across each of these pairs.

Similarly, to illustrate the model's motivation for replacing TEP with IL-76s for short movements, consider the following route: Balad Southeast, Iraq, to Ali Base, Iraq, to Kuwait International to Balad Southeast. Given the distances between these locations, we estimate that it would cost approximately $60,000 to fly an IL-76 mission across this route. The optimization model chose to fly 50 such missions, carrying on average 19.3 tons between Balad Southeast and Ali Base, 33.4 tons between Kuwait International and Balad Southeast, 4.2 tons between Ali Base and Kuwait International, and 2.3 tons between Ali Base and Balad Southeast. In contrast, we estimate that using TEP would cost approximately $65,000 to $75,000 to move slightly less cargo across just the first two origin-destination pairs.

For the examples shown here, the C-17 and IL-76 can provide more total movements at considerably less cost than can TEP. The primary reason for this is the ability of the C-17 and IL-76 to aggregate cargo across multisortie missions, supporting a mix of high-demand and low-demand origin-destination pairs on a single mission. Using TEP to provide the same set of movements would require a separate tender for each origin-destination pair, with the low-demand origin-destination pairs driving a comparatively high cost per amount moved.

A more macro-level perspective can also help explain the optimization model's preference for organic airlift over TEP. C-17 cargo moved at a cost of $1.46 per ton-mile. As discussed in Chapter Three, because we assumed a C-17 cost of $12,300 per FH and a C-17 block flying speed of 410 nmi per hour, this C-17 cost per ton-mile implies that, on average, the optimization model was able to identify an assignment of cargoes achieving a payload of 20.6 tons per C-17 flying hour.[11] Such a payload is a 32 percent improvement over the 2009 historical performance, for which the average C-17 payload was 13.9 tons per flying hour, but is still easily within the maximum payload for an intra-USCENTCOM C-17 sortie. By considering the full set of intratheater airlift requirements, the optimization model is able to assign cargo (and passengers) to aircraft in ways that make considerably more-efficient use of the organic aircraft's FH. Note that only 27 percent of actual total TEP tonnage in 2009 involved tenders

[11] The average payload is more commonly computed on a per-sortie basis. On this basis, the optimization model achieved an average C-17 payload of 13.7 tons per sortie.

that were awarded a cost less than $1.46 per ton-mile. This suggests that, if such payloads can be achieved on C-17s, under the 2009 TEP cost structure, tender is unlikely to be less expensive than organic airlift for most movements.

To further demonstrate the rationale for replacing tendered movements with organic airlift, consider as a specific example the set of tenders awarded on October 18, 2009. On this date, 45 tenders were awarded, moving a total of 566 tons of cargo a total of 297,505 ton-miles at a total cost of $1,243,135; this equates to an average of $4.18 per ton-mile. An examination of the historical GDSS data reveals that 16 of these tenders moved cargo between city pairs for which at least one C-17 or C-130 sortie was flown between October 17 and 19, 2009; for 12 of these 16 tenders, there was sufficient capacity on the corresponding C-17 or C-130 sortie to also transport the tendered cargo (assuming the maximum C-17 and C-130 cargo and passenger capacities presented earlier in this chapter). The total cost of these 12 tenders was $212,420; this entire cost, which is equal to 17 percent of the total TEP cost for this date, could have been eliminated if these cargoes had simply been added to existing organic sorties. In fact, across the entire year, 26 percent of the total tenders awarded, corresponding to 17 percent of total TEP cost, could have been moved on existing organic sorties that moved between the same city pair within one day (before or after) of the tender award.

If additional organic aircraft were available in theater, total costs could have been decreased even further. Consider the subset of tender awards appearing in Table 4.5 (none of these tenders corresponds to the set of 12 tenders that were identified previously as candidates for movement on existing C-17 or C-130 sorties).

Table 4.5
Subset of Tenders Awarded on October 18, 2009

Tender Number	Origin–Destination	TEP Cost ($)	Cargo (tons)
1	Kandahar (Afghanistan)–Camp Bastion (Afghanistan)	26,405	22.4
2	Kandahar (Afghanistan)–Camp Bastion (Afghanistan)	25,146	21.3
3	Balad (Iraq)–Kuwait City (Kuwait)	15,189	14.6
4	Al Taqaddum (Iraq)–Kuwait City (Kuwait)	25,227	14.0
5	Al Taqaddum (Iraq)–Kuwait City (Kuwait)	21,094	16.0
6	Kuwait City (Kuwait)–Balad (Iraq)	59,857	27.5
7	Kuwait City (Kuwait)–Balad (Iraq)	26,489	11.1
8	Kuwait City (Kuwait)–Balad (Iraq)	58,380	38.9
9	Bagram (Afghanistan)–Kuwait City (Kuwait)	13,107	21.8
10	Bagram (Afghanistan)–Baghdad (Iraq)	81,635	5.6
11	Kuwait City (Kuwait)–Baghdad (Iraq)	53,426	34.2
12	Kuwait City (Kuwait)–Baghdad (Iraq)	9,386	4.2
13	Al Sahra (Iraq)–Ali Base (Iraq)	35,827	24.9
14	Kuwait City (Kuwait)–Al Sahra (Iraq)	24,858	13.8
15	Kuwait City (Kuwait)–Al Asad (Iraq)	53,186	32.0

Suppose that six additional C-17s had been available in theater on this date. Table 4.6 presents a set of routes (selected from the set of potential routes available to the optimization model) that these C-17s could have flown on this date (assuming that the C-17s were correctly positioned at the start of the day). This set of C-17 routes could transport all the tendered cargoes in Table 4.5 (within the 45-ton maximum payload the model assumed for any single C-17 sortie) and generate a total savings of $329,923, which is equal to 27 percent of the total TEP cost for this date. Adding these savings to the $212,420 identified previously for tenders that could have been eliminated, the total cost to move the 566 tons of cargo that were tendered on October 18, 2009, could have been reduced by $542,343, which is equal to 44 percent of the total TEP cost for this date, if additional organic aircraft had been available in theater and if all movements were assigned to aircraft more efficiently.

An important distinction to make between this optimization model's solution and the 2009 experience is the route structure each assumes. The optimization model makes the minimum-cost allocation of cargo and passengers to aircraft missions, potentially allowing a different set of aircraft missions to be performed each day, as demand fluctuates. In USCENTCOM's actual 2009 experience, many USAF-organic aircraft flew standard theater airlift routes (STAR) missions, which Joint Publication 3-17, *Air Mobility Operations*, describes as "regularly scheduled channel missions over fixed route structures with personnel and cargo capacity available to all customers." Because C-17s and C-130s were performing STAR missions, which operate on a set schedule, the fleet of aircraft providing intra-USCENTCOM ITA in 2009 did not have the same degree of flexibility to respond to changes in demand and was thus susceptible to some degree of suboptimal performance, independent of its strategies for integrating with commercial providers.[12] It is difficult to disentangle the fraction of the optimization model's $175 million savings that is due to optimized versus STAR routing for

Table 4.6
C-17 Routes Potentially Replacing Tenders on October 18, 2009

Route	Route Distance (nmi)	Projected C-17 Cost ($)	Replacing Tenders[a]	Projected C-17 Savings ($)
Kandahar (Afghanistan)–Camp Bastion (Afghanistan)–Kandahar (Afghanistan)	170	5,100	1, 2	46,451
Balad (Iraq)–Al Taqaddum (Iraq)–Kuwait City (Kuwait)–Balad (Iraq)	724	21,750	3, 4, 5, 6, 7	126,107
Balad (Iraq)–Kuwait City (Kuwait)–Balad (Iraq)	675	20,290	8	38,090
Bagram (Afghanistan)–Kuwait City (Kuwait)–Baghdad (Iraq)–Bagram (Afghanistan)	3,476	104,420	9, 10, 11, 12	53,134
Al Sahra (Iraq)–Kuwait City (Kuwait)–Ali Base (Iraq)–Al Sahra (Iraq)	798	23,970	13, 14	36,715
Ali Al Salim (Kuwait)–Kuwait City (Kuwait)–Al Asad (Iraq)–Ali Al Salim (Kuwait)	791	23,760	15	29,426

[a] From Table 4.5.

[12] Other authors have examined this issue and found that optimized scheduling could achieve significant improvements over STAR scheduling strategies (e.g., Therrien, 2003).

USAF-organic aircraft and the fraction due to improved allocation across USAF-organic and commercial airlift options. It is likely that the additional C-17 and C-130 capacity that optimized routing makes available then substitutes for CITA across the most cost-advantageous movements. However, it is fair to claim that some of this $175 million figure could be viewed as the cost to maintain the STAR channel structure in USCENTCOM.

Optimization if Additional USAF Assets Are Made Available in USCENTCOM

Doubling the Number of C-130s Available in USCENTCOM

Additional optimization model runs increased the number of USAF-organic assets available in USCENTCOM beyond 2009 levels to determine how this would affect total delivery costs. Figure 4.2 presents the total delivery cost for the two solutions discussed previously and for a new optimization model run that increased the number of C-130s available for daily intra-USCENTCOM missions from 21 to 42 aircraft but held the number of C-17s at 15, the 2009 level. Table 4.7 then presents further details on the cost, cargo, and passengers each airlift option moved, contrasting the two optimization model solutions.

Contrasting these two solutions shows that doubling the number of C-130s that can be utilized for ITA missions reduces the total delivery cost from $1,034 million to $821 million, with a $120 million increase in C-17 and C-130 costs, which is more than offset by a $333 million reduction in CITA costs. Recall that the historical experience in 2009 had a total cost of $1,209 million; thus, the optimization model's solution with 42 C-130s is able to achieve a cost reduction of $388 million below the historical cost.

In the aggregate, the optimization model's solution has significantly increased C-130 passenger movements, with a corresponding decrease in C-17 passenger movements; it has also

Figure 4.2
Total Delivery Cost with Double the Number of C-130s and the Same Number of C-17s, Versus 2009 Levels

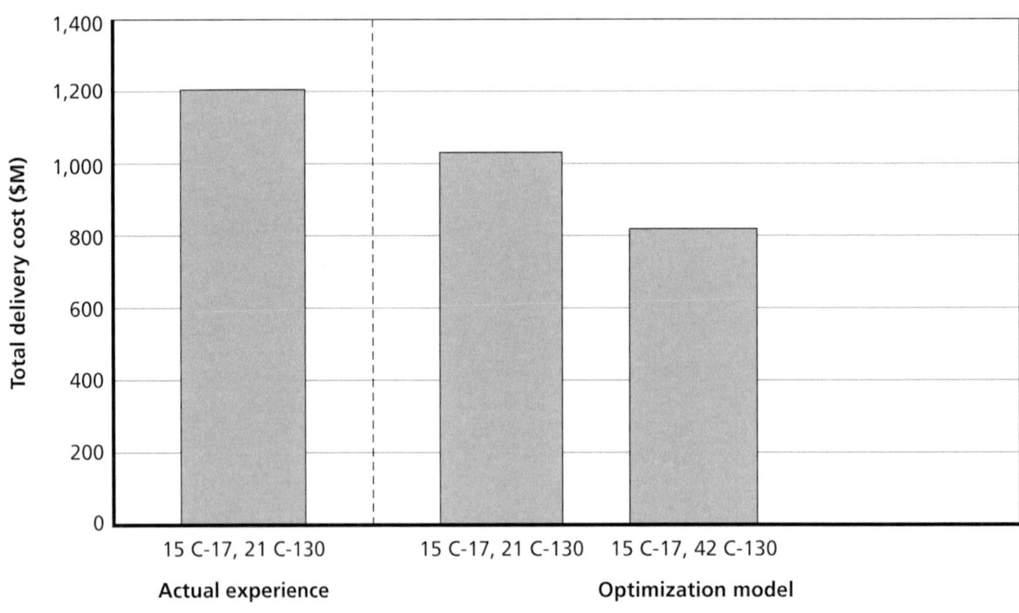

RAND *TR1313-4.2*

Table 4.7
Comparison of Optimization Model Solutions Utilizing Varying Numbers of C-130s

Airlift Type	Optimization Solution, with 21 C-130s and 15 C-17s					Optimization Solution, with 42 C-130s and 15 C-17s				
	Cargo (000 tons)	Passengers (000s)	Cargo Ton-Miles (M)	Passenger-Miles (M)	Cost ($M)	Cargo (000 tons)	Passengers (000s)	Cargo Ton-Miles (M)	Passenger-Miles (M)	Cost ($M)
C-17	183	621	207	411	303	237	380	282	304	349
C-130	43	512	19	209	157	85	755	32	296	231
TEP	41	—	43	—	182	22	—	26	—	100
IL-76	137	—	85	—	392	60	—	17	—	141

significantly increased the total amount of cargo moved on C-17s and C-130s while greatly decreasing the cargo moved on TEP and IL-76 charters.

As before, examining the movements at a finer level of detail offers greater insights. When the number of employed C-130s was doubled, 241,000 passenger movements were reassigned from C-17s to C-130s, and 72 percent of these reassigned movements were relatively short, between Kuwait and Iraq (in both directions). As an illustration of why the model makes such reassignments, consider the route between Ali Al Salim, Kuwait, to Balad Southeast, Iraq, and back to Ali Al Salim. Using differences in each aircraft's flying speed and our cost per flying hour calculations, we estimated that it costs $19,000 to fly a single C-17 across this route and $16,000 for a single C-130. In the optimization result with 21 C-130s, C-17s performed 173 such missions and transported 41,000 passengers, while C-130s performed 105 such missions and transported 13,000 passengers. The total cost for all these missions was $5 million. Contrast this with the optimization model results for 42 C-130s, for which the model assigned only 66 such C-17 missions, transporting 15,000 passengers, while C-130s are assigned 354 such missions, transporting 45,000 passengers; the total cost has increased to $7 million. Note that the total passengers moved are comparable, but not equal, across the two optimization results because passengers could have been moved between the same city pairs along different routes. Thus, for a comparable level of passenger movements, the model has selected missions that generate a 40-percent cost increase. Why would this be desirable?

When the number of employed C-130s was doubled, the total tons of cargo transported by C-17 increased by 54,000 tons; the total tons of cargo transported by C-130 increased by 42,000 tons; and total tons of cargo transported by IL-76 decreased by 78,000 tons. For C-17s, 78 percent of this increase occurred across relatively long routes into Afghanistan from Kuwait, Qatar, Iraq, and Kyrgyzstan; 46 percent of the reduction in IL-76 cargo occurred over these same origin-destination pairs. Across relatively short routes, 43 percent of the C-130 increase and 21 percent of the IL-76 decrease occurred across Kuwait–Iraq or intra-Iraq routes, while 36 percent of the C-130 increase and 17 percent of the IL-76 decrease occurred across intra-Afghanistan movements.

The optimization model is utilizing the additional 21 C-130s in two roles: replacing IL-76s for short movements and replacing C-17s for passenger movements so that the released C-17s can then be used to displace IL-76s for long sorties. The primary motivation for such a shift is that the IL-76 costs much more than the C-17 for a comparable speed and cargo payload capacity.[13] According to the Silk Way Airlines FY 2010 IL-76 contract, the cost to operate a "live" (i.e., cargo-carrying, as opposed to empty aircraft positioning) segment is $89 per nautical mile.[14] Recall that the TWCF DoD SAAM C-17 cost is $12,317 per flying hour; dividing that cost by a 410 nmi per hour C-17 planning factor yields a C-17 cost of $30 per nautical mile, one-third the IL-76 cost.[15] This provides an incentive to reassign the longest, and thus costliest, routes from the IL-76s and to the C-17s and explains why the model is willing

[13] AFPAM 10-1403, 2003, provides C-17 planning factors of 45 tons payload per sortie and a block speed of 410 nmi per hour. Skyline Aviation, Ltd., undated, suggests an IL-76 maximum payload of 44 tons per sortie and 410 nmi per hour.

[14] This contract provides fixed costs for movements between a list of 11 locations in Iraq, five locations in Afghanistan, one location in Qatar, and one location in Kuwait, based on the countries of origin and destination; this $89 per nautical mile rate applies to "alternative locations within" USCENTCOM.

[15] Note that this IL-76 cost is much larger than the FY 2010 TWCF DoD SAAM rate for commercial augmentation aircraft, whose most-expensive cargo rate was $1.09 per ton-mile for one-way transport on medium class–body aircraft (e.g.,

to accept increased costs in one area (passenger movements via C-130) to generate even larger cost reductions in another area.

Considering the increase in C-130 cargo movements, a similar calculation, dividing our estimated $6,800 per FH by a 270 nautical mile per hour planning factor, produces a rate of $25 per nautical mile. This comparison is, however, more complicated because the IL-76 can carry considerably more cargo per sortie than the 12 tons per sortie planning factor for C-130s. In the optimization model results for 21 C-130s, the average payload per IL-76 sortie was 18.5 tons, excluding empty positioning and depositioning sorties, suggesting that many IL-76s were not flying at maximum capacity, affording opportunities for the lower-cost-per-FH C-130 to perform some of these movements when the number of employable C-130s was doubled.

These calculations do not, however, include the cost to deploy additional aircraft. Thus, were it possible to double the number of C-130s in theater for less than $213 million,[16] the strategy could have been cost-effective for meeting USCENTCOM ITA demands in 2009.[17] We did not include the cost to deploy additional aircraft in this analysis because any estimate of this cost varies significantly depending on such factors as the specific deployment location. If the additional aircraft were deployed to a location that already supports C-130 operations, the deployment costs would be significantly less than if the aircraft were deployed to a new location that did not previously support C-130s. The deployment costs also vary depending on the duration for which individual aircraft and individual servicemembers are deployed and also depending on whether the deployed personnel are active-duty or ARC members because activating some ARC positions incurs additional expenses.

However, to give an impression of the order of magnitude of these deployment costs, consider a case in which eight C-130s were deployed to a location that was already supporting 20 C-130s. Based on a separate RAND model that estimates the total manpower necessary at a deployed location,[18] an additional 312 positions would need to be deployed to this location. Of these, 236 would be tactical airlift aviation or maintenance positions, with the remainder for various base support functions. Filling all these positions with active-duty personnel would potentially also incur additional expenses, for a family separation allowance ($250 per position per month), hardship duty pay (which varies by the location of deployment; in Kuwait, it is $100 per position per month), and hostile fire/imminent danger pay ($225 per position per month). If this deployment were to be sustained for an entire year, these additional expenses would total $2,152,000.

Assuming that the aircraft would remain deployed for six months, it would be necessary to fly the aircraft between their home station and the deployed location. Assuming a notional deployment between Dover Air Force Base (AFB) and Ali Al Salim, Kuwait (a flight of 5,603 nmi and 20.8 FH each way, at a planning factor C-130 speed of 270 nmi per hour), at $6,800

B767-200F), even at a fully loaded IL-76 cargo weight of 44 tons, this TWCF rate of $48 per nautical mile is slightly more than one-half of the $89 per nautical mile rate for IL-76s.

[16] This $213 million is the savings the optimization model obtains with 42 C-130s over the optimization solution with 21 C-130s, not the total $388 million in savings generated by the optimization model with 42 C-130s over the historical experience.

[17] We performed additional optimization model runs with even more C-130s (e.g., 15 C-17s and 76 C-130s), but the model found relatively little benefit to employing more than 42 C-130s (i.e., the optimized total cost did not decrease significantly when more than 42 C-130s were made available).

[18] The Strategic Tool for the Analysis of Required Transportation (START) model, see Snyder and Mills (2004).

per FH and two round trips per year, the total annual cost would be $4,515,000. Finally, if we assume that each individual would remain deployed for six months, it would be necessary to transport the individuals between their home station and the deployed location. Assuming the same notional locations, the FY 2010 DoD Channel Passenger Tariff rate would cost a total of $1,834 per person between Dover AFB and Kuwait City, Kuwait.[19] Across an entire year, this would total $2,289,000, which is likely an overestimate because some of these personnel could potentially be transported on the deploying C-130 aircraft.

The sum total of these annual costs per squadron of eight C-130s deployed to an established C-130 base, with both aircraft and active-duty personnel deploying for intervals of six months apiece, is $8,957,000.[20] Other costs are associated with deployment, such as U.S. government payments to host-nation contractors to provide aspects of base support, but because we could not identify a means of estimating these costs on a marginal basis (that is, per additional squadron deployed), we did not include them here. Given this estimate of additional deployment costs, the $27 million necessary to deploy three additional squadrons of C-130s is much less than the $213 million reduction in ITA delivery costs that our optimization model suggests could be achieved were the number of deployed aircraft doubled. Thus, these additional deployment costs, while not insignificant, would be much more than offset by the potential reductions in ITA delivery costs.

Doubling the Number of C-17s Available in USCENTCOM

In another optimization model run, we increased the number of C-17s available for daily intra-USCENTCOM missions from 15 to 30 aircraft but held the number of C-130s at 21, its 2009 level. Figure 4.3 presents the total delivery cost for this new model run, along with the three solutions discussed previously. Table 4.8 then presents further details on the cost, cargo, and passengers moved by each airlift option, contrasting the optimization model solutions that differ only in the number of C-17s available for ITA movements.

Contrasting these two solutions shows that doubling the number of C-17s available for ITA missions reduces the total delivery cost from $1,034 million to $703 million, with a $113 million increase in C-17 and C-130 costs that is more than offset by a $444 million reduction in CITA costs. Recall that the historical experience in 2009 had a total cost of $1,209 million; thus, the optimization model solution with 30 C-17s is able to achieve a cost reduction of $506 million below the historical cost. In the aggregate, this solution has significantly increased the passenger and cargo movements using C-17s, with a corresponding decrease in C-130 passenger movements, and has significantly decreased cargo movements on TEP and IL-76 charters. This is due to the C-17's relative cost advantages, which we discussed previously. The C-17 can perform many cargo movements at lower cost than either CITA alternative. Moreover, it can perform many passenger movements at less cost than the C-130; unlike the previous optimization run in which the number of C-130s was doubled but the number of C-17s held constant, there is no need to accept increased passenger movement costs via C-130 to free C-17s for cargo movements here.

---

[19] The Channel Passenger Tariff is stated in terms of $1,733 per person between Dover AFB and Baghdad International, Iraq, and an additional $101 between Baghdad, Iraq, and Kuwait City, Kuwait.

[20] For comparison, the START model estimates a requirement of 978 total deployed positions for deploying the same aircraft to a bare base. The comparable total cost associated with these 978 positions is $18,438,000.

Figure 4.3
Total Delivery Cost with Double the Number C-17s and the Same Number of C-130s, Versus 2009 Levels

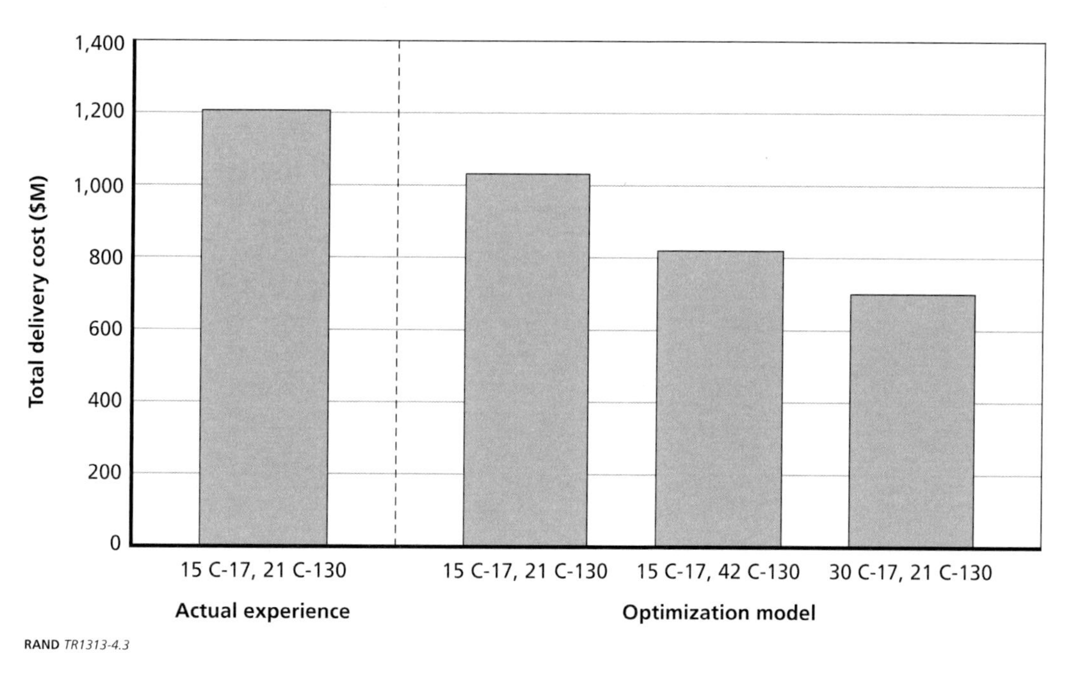

RAND *TR1313-4.3*

In fact, the $703 million cost that was achieved with 30 C-17s was essentially equal to the global minimum cost that could be achieved if we allowed the model access to an unlimited pool of C-130s and C-17s. What this suggests is that there would be no benefit to deploying more aircraft than required to employ 30 C-17s and 21 C-130s each day, in this optimization model run. In this solution, the use of TEP and IL-76 over time shows a relatively constant use of CITA over the entire year, suggesting that CITA is not being used solely for surge capacity on days with unusually high volume. Rather, the 15,000 and 23,000 tons that TEP and the IL-76s, respectively, deliver in this solution are the cargoes for which these two alternatives are the most cost-effective delivery options.[21] This suggests that, even though USAF aircraft appear to be more cost-effective for most movements, CITA options should be retained for some small fraction of USCENTCOM ITA movements.[22]

Sensitivity Analyses

To ensure that these findings were not overly dependent on our specific cost estimates, we performed sensitivity analyses that varied the cost structures for the various airlift options. Because our analyses found TEP to be less-preferred than any of the other airlift options, we

---

[21] Recall our earlier assumption that limited the optimization model's set of allowable sorties for all IL-76s and USAF aircraft to the sorties that were actually flown by C-130s and C-17s in USCENTCOM in 2009. Because 6,400 tons were moved between origin-destination pairs that only TEP served in 2009, TEP was the only delivery option available to the model for such movements. These 6,400 tons include TEP movements to Djibouti (in USAFRICOM) that were included in our set of data movement requirements but were only allowed to be performed via TEP.

[22] This assumes that CITA providers would still elect to participate at the reduced volumes.

Table 4.8
Comparison of Optimization Model Solutions Utilizing Varying Numbers of C-17s

	Optimized, with 21 C-130s and 15 C-17s					Optimized, with 21 C-130s and 30 C-17s				
Airlift Type	Cargo (000 tons)	Passengers (000s)	Cargo Ton-Miles (M)	Passenger-Miles (M)	Cost ($M)	Cargo (000 tons)	Passengers (000s)	Cargo Ton-Miles (M)	Passenger-Miles (M)	Cost ($M)
C-17	183	621	207	411	303	325	803	319	469	444
C-130	43	512	19	209	157	41	332	16	134	129
TEP	41	—	43	—	182	15	—	18	—	79
IL-76	137	—	85	—	392	23	—	4	—	51

performed three additional optimization model runs in which 15 C-17s and 48 C-130s were available for ITA movements.[23] Table 4.9 presents the cargo each airlift option moved in the historical data and under three optimization model runs, one with our baseline cost estimates, one in which TEP costs were scaled by 9/10 and other costs by 10/9, and one in which TEP costs were scaled by a factor of 4/5 and other costs by a 5/4 factor.

In these sensitivity analysis outputs, the amount of cargo transported via C-17 and C-130 does not significantly change as their per-hour flying costs increase by up to 25 percent. This suggests that our finding that USAF aircraft are more cost-effective than TEP for most 2009 ITA movements within USCENTCOM is fairly robust. Note, however, that the relative preference for IL-76 charters over TEP is highly dependent on the specific cost structure assumed for each option. As IL-76 costs increase and TEP costs decrease, the amount of cargo transported via TEP increases, with most of this increase coming at the expense of cargo previously transported via IL-76. This suggests that, while IL-76s may be slightly more cost-effective than TEP for most of the 2009 ITA movements within USCENTCOM best-suited for CITA, this preference is not robust to moderately sized changes to the relative IL-76 and TEP cost structures.

Table 4.9
Sensitivity Analysis

	Cargo Moved (000 tons)			
		Optimization Model Results, with 48 C-130s		
Airlift Type	2009 Experience (daily average 21 C-130s)	Best Estimates of Costs	TEP Costs Scaled by 9/10, Other Costs Scaled by 10/9	TEP Costs Scaled by 4/5, Other Costs Scaled by 5/4
C-17	173	239	235	232
C-130	54	87	88	83
TEP	133	21	34	50
IL-76	38	57	47	38

[23] This sensitivity analysis should not be used to contrast the 48 C-130 results with those for the previous optimizations; rather, it should be used to contrast the optimization model's relative use of each mode (with the constant 48 C-130s available) as the underlying cost structure is modified.

CHAPTER FIVE

Conclusions and Potential Extensions to Research

The analysis presented in this report examines the cost-effectiveness of DoD's use of CITA within USCENTCOM. We developed a set of models to determine the cost of obtaining ITA from commercial sources and to identify the marginal cost associated with increasing the deployed FH performed by USAF aircraft. An optimization model was then built to identify the minimum ITA cost that could be achieved given a set of airlift resources. We applied this analytic framework to the set of ITA cargo and passenger movements that were performed in USCENTCOM in 2009, and then examined the model outputs.

Conclusions

Our examination of the demands for ITA and the costs that were paid to CITA providers in USCENTCOM in 2009 led to two general conclusions.

C-17 and C-130 are both generally more cost-effective than CITA, but CITA options should be retained to supplement USAF aircraft. Across all optimization model runs, the model demonstrated a clear preference for increasing the use of USAF aircraft and decreasing the use of CITA, although there was a relatively small fraction of the USCENTCOM ITA demands for which IL-76 charters and TEP tenders were the most cost-effective options.

For a level of USAF resources equal to what was historically used in USCENTCOM (15 C-17s and 21 C-130s available each day for ITA movements), the optimized allocation of cargo and passengers to airlift options was able to reduce costs by $175 million from the historical performance. This allocation replaced TEP with C-17s for long movements and TEP with IL-76s for short movements. The optimization model made these changes primarily because of the ability of the C-17 and IL-76 to aggregate cargo across multisortie missions, supporting a mix of high-demand and low-demand origin-destination pairs on a single mission. Using TEP for the same sets of movements would require a separate tender for each origin-destination pair, with the low-demand origin-destination pairs driving a comparatively high cost per amount moved.

With double the number of C-130 aircraft available to support USCENTCOM ITA but the same number of C-17s, the optimized allocation could reduce costs by slightly more than $210 million from the optimization result for 21 C-130s (for a total savings of approximately $390 million). This allocation would do so by moving significantly more passengers on C-130s (with correspondingly fewer passengers on C-17s), replacing IL-76s with C-17s for long movements, and replacing IL-76s with C-130s for short movements. Although transferring passengers from C-17s to C-130s would often increase the costs of performing such movements, the model elected to do so because such a strategy would free C-17s to take over long-distance cargo movements from the IL-76. This strategy was cost-effective because, on a per-mile cost basis, the IL-76 costs three times as much as the C-17 for a comparable aircraft block speed and payload. The model elected to use some of the remaining C-130 increase to replace IL-76s across short-distance movements, likely because the average IL-76 payload for the optimization model run with 21 C-130s was only 18.5 tons, suggesting that many IL-76 sorties used a larger and relatively expensive aircraft to move small, C-130-sized payloads.

With double the C-17s available to support USCENTCOM ITA but the same number of C-130s as the historical level, the optimized allocation could reduce costs by slightly more than $330 million from the optimization result for 15 C-17s (for a total savings of approximately $510 million). This allocation would do so by moving significantly more passengers and cargo on C-17s, with corresponding decreases in C-130 passenger movements and cargo movements using TEP and IL-76 charters. This is due to the C-17's relative cost advantages, which we discussed earlier.

These cost reductions do not account for the increased costs to the USAF associated with deploying these additional aircraft to USCENTCOM, but our preliminary analysis suggests that these added costs would be much smaller than the potential savings.

The minimum cost, which was achieved with 30 C-17s, was essentially equal to the global minimum cost that could be achieved if we allowed the model access to an unlimited pool of C-130s and C-17s. The model solution with 30 C-17s used TEP and IL-76s to transport 4 and 6 percent, respectively, of the total cargo tonnage. For these cargoes, therefore, CITA appears to be the most cost-effective delivery option, suggesting that, even though USAF aircraft appear to be more cost-effective for most of the USCENTCOM ITA movements, CITA options should be retained for some small fraction of movements.[1]

We also conducted sensitivity analyses, running the optimization model with lower per-movement costs for TEP and higher costs for C-130s, C-17s, and IL-76s. In these sensitivity analyses, the amount of cargo transported via C-17 and C-130 did not significantly change as their per-hour flying costs increased by either 11 or 25 percent, suggesting that our finding that USAF aircraft are more cost-effective than commercial options for most 2009 ITA movements within USCENTCOM is fairly robust. However, as IL-76 costs increased and TEP costs decreased, the amount of cargo transported via TEP increased, with most of this increase coming at the expense of cargo previously transported via IL-76. This suggests that, while IL-76s may be slightly more cost-effective than TEP for most of the ITA movements within USCENTCOM best-suited for CITA, this preference is not robust to moderately sized changes to the relative IL-76 and TEP cost structures

Decision-support tools are needed to assist the CAOC AMD and USCENTCOM Deployment and Distribution Operations Center with daily airlift cargo allocation decisions. When multiple airlift options exist for any specific movement, identifying the cost associated with each airlift option is only the first step of a cost-effectiveness evaluation. Given a large collection of movement requirements and a set of airlift alternatives, it is necessary to solve a routing problem and an assignment problem: which movements to assign to which missions. At the time of this analysis, the CAOC AMD lacked sophisticated decision-support tools to assist in its daily cargo-aircraft allocation decisions. The extremely large number of potential

[1] This assumes that CITA providers would still elect to participate at the reduced volumes.

assignments prohibits any individual from considering all feasible options and selecting the most effective solution without the aid of a computer model. We developed an optimization model to perform such movement-to-mission assignments and found that the model was able to identify significant improvements to the historical performance. USCENTCOM actually incurred approximately $1,210 million in total ITA delivery costs; the model found a solution that could have reduced this cost by up to $175 million without increasing the daily number of employed C-17s and C-130s in theater. This suggests that an investment in the development of such tools for AMD could achieve large savings.

Potential Extensions to Future Research

As mentioned above, we were not able to thoroughly analyze the total marginal cost to the USAF to deploy a C-130 or C-17 squadron. Many variables significantly influence such costs, such as whether the squadron is an active duty, ARC, or associate unit; whether the squadron is deploying to a location that already supports a deployment of similar aircraft; and the arrangements between the U.S. government and host-nation contractors to provide base support. Further analysis could examine this issue in more detail and provide a range of estimates for use in mobility and beddown planning.

In our discussion of the optimization model results, we discussed a shortcoming of the model with respect to the effects of demand uncertainty. On peak demand days, the model's myopic perspective attempts to satisfy this very large demand at minimum cost, even if this requires chartering a large number of aircraft that may be underutilized in the future. It does so even if paying more in the near term to move this atypically large demand via a tender program would avoiding the long-term drawbacks associated with a large charter purchase. We were able to overcome this limitation in our analysis of 2009 experience by adding a constraint to the model that limited the maximum number of aircraft that could be chartered. The problem in practice is that it is not clear exactly what constitutes a given day's demand as being so unusually large as to be unlikely to be encountered again. What is needed is a model formulation that looks back across some recent period; determines how unusual a day's demand is with respect to both recent history and some knowledge of expected future operations; and then identifies the risk, in terms of exposure to potential future cost, associated with entering into a lower-cost but less-flexible agreement, such as aircraft chartering, rather than a higher-cost but more-flexible arrangement, such as tendering, all while recognizing the uncertainty associated with future levels of demand. Further analysis could extend our optimization models to account for these effects of demand uncertainty.

APPENDIX A

Data Merging

To determine the historical level of CITA utilization within USCENTCOM in 2009, it was necessary to merge cargo and passenger movement data from multiple data systems. This appendix describes the process of generating the two new data sets for this analysis, which we will call the "requirements" and "execution" data sets.

We used three primary data sources to create these new merged data sets:

- **GATES** collects data on the set of cargo and passengers that pass through Air Force aerial ports. From it, we obtained a data pull covering all of calendar year 2009.
- **GDSS** is a command and control system for the dissemination of airlift and tanker mission plans for all mobility air force operations. From it, we obtained a data pull covering all of calendar year 2009.
- The **TEP data set** is not a formal data system but rather a spreadsheet that USTRANSCOM's Acquisition Directorate provided us containing TEP data covering October 1, 2008, through December 17, 2009. This data set contained information on all tenders that were offered to commercial carriers; the bids that were received from each of the seven carriers that participated in TEP at that time, along with an identification of the winning bid; and pallet-level detail on the movement history of all items moved through TEP.

GATES and GDSS accumulate data for both USAF-organic aircraft and the chartered commercial cargo aircraft AMC controls. The GATES data set provides pallet and passenger information, such as origin and destination, date-time, and size (weight, height, and volume). The GDSS data set provides sortie-level information, such as aircraft type, departure and arrival locations, and date-time. The GDSS data set can be viewed as assessing how the Air Force executed the airlift requirement reported in GATES.

Both GATES and GDSS contain a common data field identifying the mission (AMC Mission ID), which allowed us to link the two data sets. Each mission identification corresponds to a series of sorties, or a route, that can be reconstructed from GDSS. Thus, we could check to see whether any segment of a given route consisted of an intra-USCENTCOM sortie.

We began by eliminating from further consideration (a) all GDSS records whose mission IDs did not correspond with any GATES records and (b) all GATES records whose mission IDs did not correspond with any GDSS records. We then examined all GDSS sorties corresponding to each remaining mission ID. If all sorties within a mission had both origin and destination in USCENTCOM, the corresponding GATES record was added to the "requirements" data set, and all sorties in GDSS corresponding to this mission ID were added to the "execution" data set. If a mission included an intra-USCENTCOM sortie and at least one sortie with either its origin or destination outside USCENTCOM, it was necessary to identify the intratheater segment of this mission. The execution data set included only GDSS sorties corresponding to the intra-USCENTCOM segment of the mission. For such missions, we identified a "pseudoorigin" and "pseudodestination" corresponding to the intratheater segment of the mission. The requirements data set included all GATES records corresponding to this mission ID, with pseudoorigins and -destinations replacing their actual origins and destinations.

We computed the cargo and passenger loads for each sortie in the execution data set from the merged data. For each sortie in the execution data set, we summed the weight, pallet positions, and passenger counts for all requirements data set records that would have flown on this sortie. Note that some of these "execution" records could be empty, i.e., show no cargo or passengers, if the sortie had a mission ID corresponding to a requirements data set record but occurred prior to the loading of the first (or following the unloading of the last) intra-USCENTCOM cargo or passenger corresponding that mission ID.

Observe that, by linking the set of cargo and passengers from GATES with the set of sorties from GDSS, the requirements data set includes all cargo and passengers that moved on at least one intra-USCENTCOM sortie. Note that our data set potentially includes some cargo and passengers that were simply transiting USCENTCOM while moving between two other theaters. As a practical matter, this was likely rare and should thus should account for very little cargo or passengers.

Because TEP serves only USCENTCOM locations, with the exclusion of Djibouti (in USAFRICOM), and since the TEP data do not contain any information regarding the sorties that accomplished these movements, we simply added all TEP movements to the requirements data set.[1]

The requirements data set contains a total of 404,000 tons of cargo. Over 98 percent of this cargo was moved by C-17, C-130, IL-76, AN-124 or TEP. Other aircraft transported 6,700 tons in this data set, of which C-5s moved 94 percent. The requirements data set also contained a total of 1,134,000 passenger movements. Because, as discussed in Chapter Three, our models required all passengers to travel on C-17 or C-130 aircraft, this data set included only intra-USCENTCOM passenger movements that occurred on C-17s or C-130s. This allowed us to exclude the movements of very small numbers of passengers on small aircraft (such as C-21A Learjets) that are not viable candidates for C-17 or C-130 movement.

---

[1] Our requirements data set included these TEP movements to Djibouti. We excluded some records in the TEP data set that were identified as having never completed delivery.

APPENDIX B

Evaluation of the Theater Express Program

DoD's TEP solicits commercial carriers to bid in a daily spot market to move rolling stock and palletized cargo within USCENTCOM.[1] It is designed to augment military and commercial contract lift activities. Since the program began in mid-2006, it has become an increasingly important aspect of intratheater logistics activities within USCENTCOM. This appendix provides background on the TEP and presents our statistical analysis seeking a better understanding of how factors affect the cost of using the program. This analysis relies on program data covering October 1, 2008, to December 17, 2009.

Competition in TEP Auctions

Programs like TEP rely on competition to discipline market participants. That is, a carrier would like to make excess profits on its business activities in TEP by bidding high, but the risk of losing business to other carriers keeps its bids in line with costs. On average, 4.2 carriers bid on each TEP tender, suggesting a healthy amount of competition among carriers in the program. Only 9 percent of all tenders have one bidder. Because carriers do not know which tenders their competitors are bidding on, there are pressures to place competitive bids even in these cases. Figure B.1 shows the distribution of the number of carriers that bid on TEP tenders.

Origins and Destination of Shipments Through the TEP

Table B.1 presents a summary of TEP movements categorized by country of origin and destination. Approximately one-third of all TEP shipments originated from the distribution depot in Kuwait, whether measured in terms of tons of cargo shipped or in terms of value. The depot in Kuwait is a commercial facility located 15 km north of Camp Arifjan in Kuwait and in close proximity to Kuwait City International Airport and the Shuwaik commercial ocean port. While the distribution depot in Kuwait initially focused on servicing units located in Kuwait and Iraq, its role has expanded to include the entire USCENTCOM AOR (see Newton and Turnage, 2007).

[1] TEP cannot be used to transport explosives; cargo requiring ventilation (e.g., liquid oxygen carts); wet or dry ice; registered mail; cargo requiring an escort, courier, or signature and tally record; or personnel.

Figure B.1
Distribution of Number of Bidders in TEP Tender Auctions

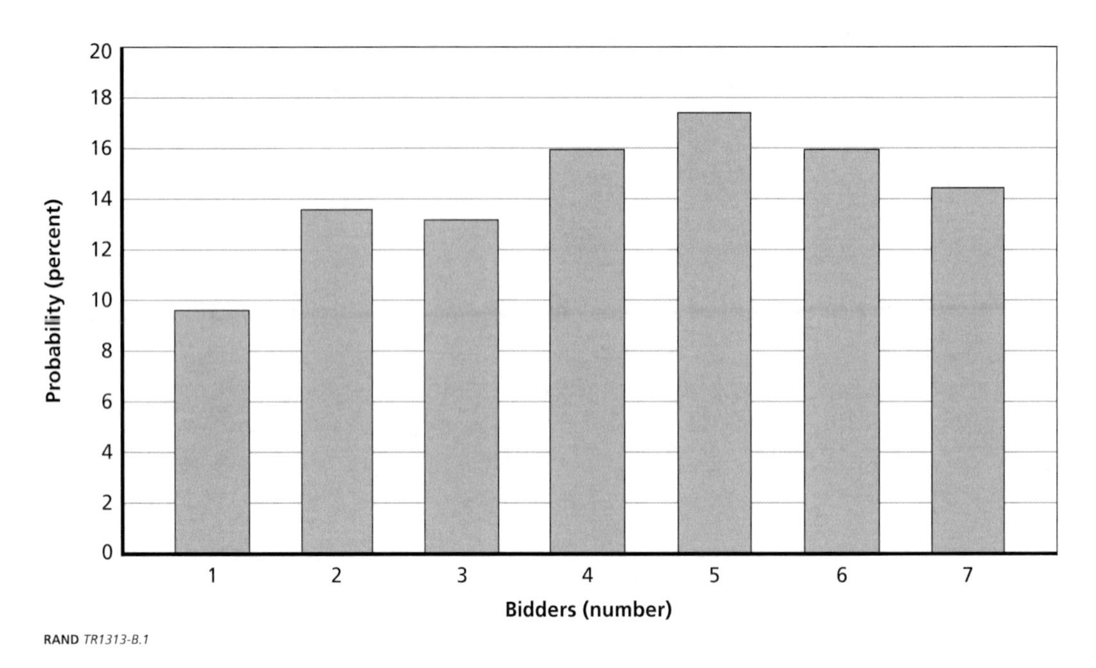

RAND *TR1313-B.1*

Table B.1
Summary of Theater Express Movements by Country of Origin and Destination (October 1, 2008, to December 17, 2009)

	Outgoing Movements			Incoming Movements		
Country	Number of Awards	Tons	Shipment Costs ($M)	Number of Awards	Tons	Shipment Costs ($M)
Distribution depot, Kuwait	5,320	64,547.8	163.7	0	0.0	0.0
Djibouti[a]	0	0.0	0.0	116	674.9	4.8
Afghanistan	1,928	19,064.8	40.8	5,640	69,614.2	245.1
Bahrain	236	1,325.6	6.6	230	703.6	3.0
Kuwait	979	7,176.0	29.6	1,576	15,169.5	28.2
United Arab Emirates	2	20.3	0.2	182	396.7	2.1
Oman	88	905.9	6.2	89	383.5	2.8
Pakistan	0	0.0	0.0	1	2.8	0.0
Iraq	8,069	53,028.3	145.0	8,910	66,134.7	138.6
Qatar	1,069	11,079.5	47.4	811	3,134.7	9.6
Kyrgyzstan	36	317.8	1.7	171	1,241.2	7.1
Turkmenistan	0	0.0	0.0	1	10.2	0.1
Total	17,727	157,466.0	441.3	17,727	157,466.0	441.3

[a] Djibouti falls within USAFRICOM. It is the only origin-destination serviced by TEP that falls outside of USCENTCOM.

Iraq, as an origin, accounted for approximately another one-third of all cargo shipments between October 1, 2008, and December 17, 2009. These shipments where spread over a number of different airports located within the country. Qatar and Afghanistan rank next as important origin countries.

Not surprisingly, the primary destinations of cargo transported through TEP were in Afghanistan and Iraq. Approximately 44 percent of all cargo, measured by weight, arrived in Afghanistan, while Iraq received approximately 42 percent of all cargo. Shipments to Afghanistan and Iraq accounted for approximately 87 percent of the program's cost.

TEP Activity over Time

Use of TEP has varied somewhat over the period for which we have data. Figures B.2, B.3, and B.4 suggest that tenders were increasing through the summer of 2009 but declined somewhat over the latter half of 2009. Plots of the number of tons moved and tender costs also suggest a ramp-up in the program between October 2008 and summer 2009, but the decline after then is small relative to the reduction in the number of tenders, suggesting that tenders were getting larger in terms of both tons moved and costs.

Analysis of Drivers of Tender Costs

The tender data show considerable variation in the bids received. In general, we would expect carriers to vary their bids based on the amount of cargo moved and the flight times. Furthermore, as the number of bidders increases, the winning bid is likely to be lower because

Figure B.2
Monthly Tender Awards

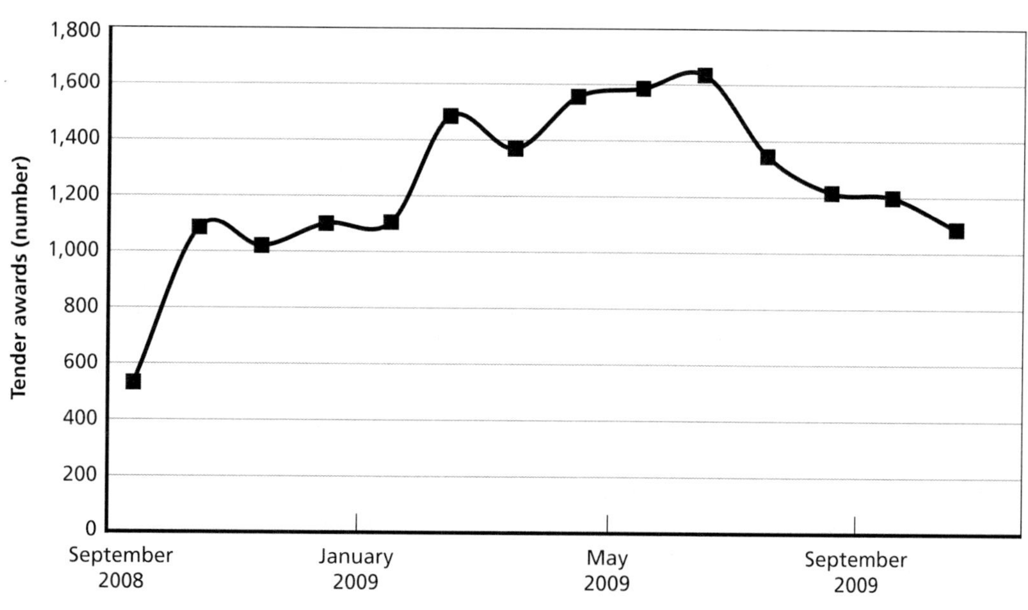

RAND *TR1313-B.2*

Figure B.3
Monthly Tons Moved

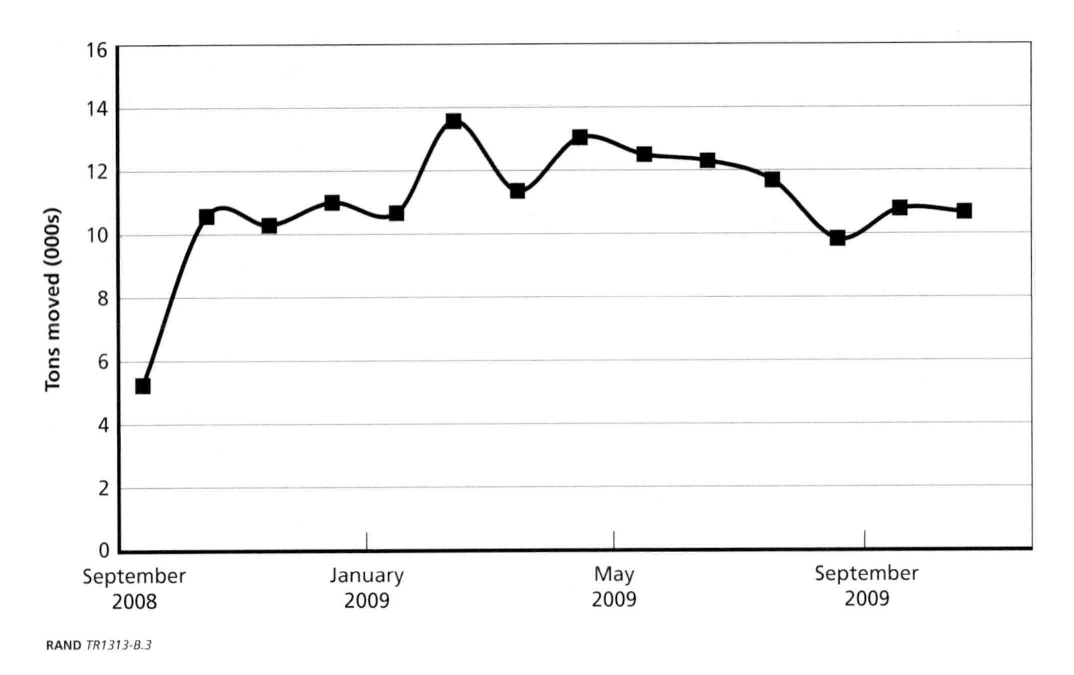

RAND *TR1313-B.3*

Figure B.4
Monthly Tender Cost

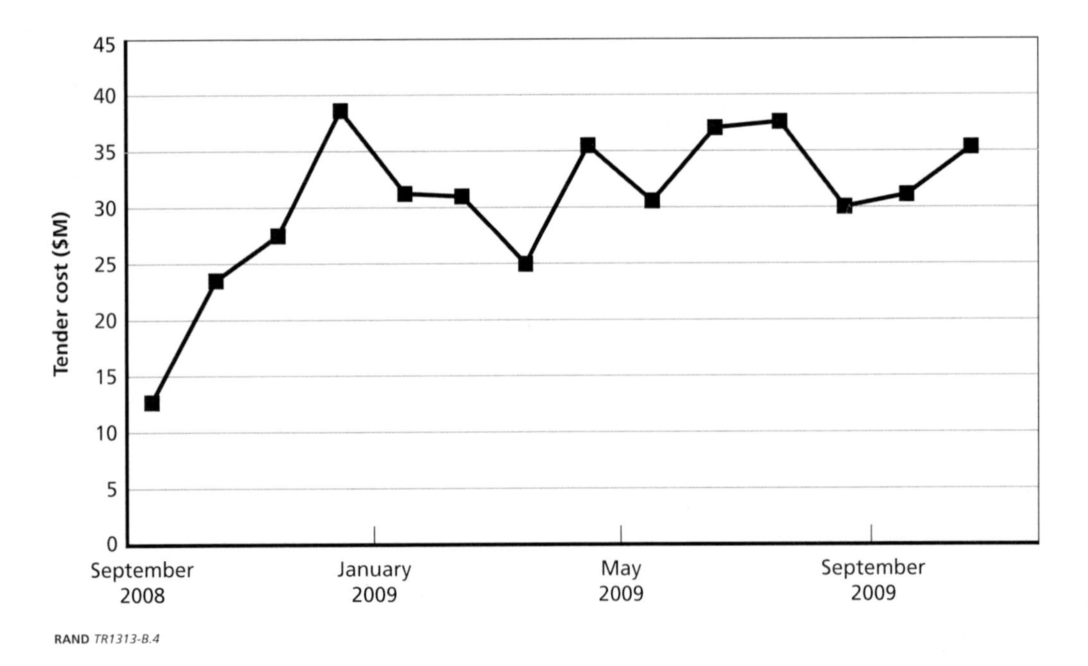

RAND *TR1313-B.4*

the administrator has more bids to select among and because competitive pressures may be stronger on that route. There may also be systematic differences between origins and destinations that make them more or less desirable to certain carriers. In the statistical analysis, we attempted to control for these factors, as well as for seasonal effects. Specifically, we model the tendered costs per pound of cargo (p_{odt}) as being a function of

- flying time (t_{od}), measured in the equivalent required flying time of a C-130
- quantity of cargo shipped (q_{odt}), measured in tons of cargo transported
- level of competition (N_{odt}), measured in terms of the number of bidders
- seasonal effects (γ_m), captured by monthly dummy variables
- origin and destination characteristics, measured using either origin (η_o) and destination (μ_d) fixed effects, or using route fixed effects (λ_{od}).

Here, the subscript o denotes origin; d denotes destination; t denotes time measured in calendar days; and m is an index on months with each t falling in one month m. Using data from the TEP, we ran regressions of the form

$$\log(p_{odt}) = \alpha + \beta\log(t_{od}) + \delta\log(q_{odt}) + \phi\log(N_{odt}) + \gamma_m + \eta_o + \mu_d + \varepsilon_{odt}.$$

Here, ε_{odt} is an independent and identically normally distributed random error term. We also ran the regression without the monthly and origin and destination fixed effects. The coefficients of the logged variables can be interpreted as elasticity estimates.

In a second specification, we used route origin-destination dummy variables (λ_{od}) to control for shipping distance and origin-destination characteristics instead of t_{od}, η_o, and μ_d. Specifically, in the secondary specification, we estimated a regression of the form

$$\log(p_{odt}) = \alpha + \delta\log(q_{odt}) + \phi\log(N_{odt}) + \gamma_m + \lambda_{od} + \varepsilon_{odt}.$$

Table B.2 presents the results of these regressions. These results suggest the findings described in the following subsections.

Table B.2
Regression Results for Tender Cost Regressions

Model	1a	1b	2
Dependent variable		$\log(p_{odt})$	
β (coef. on $\log(t_{od})$)	0.415**	0.297**	NA
Standard error	0.006	0.008	NA
δ (coef. on $\log(q_{odt})$)	–0.266**	–0.282**	–0.296**
Standard error	0.004	0.004	0.004
ϕ (coef. on $\log(N_{odt})$)	–0.554**	–0.475**	–0.383**
Standard error	0.008	0.009	0.009
Month dummies (γ_m)	No	Yes	Yes
Origin and destination dummies (η_o, μ_o, or λ_{od})	No	Yes	Yes
Observations	17,618	17,618	17,725
Parameters	4	73	325
R^2	0.522	0.622	0.681

NOTE: Within this table, * indicates the coefficient is different from 0 with at least 95 percent confidence, and ** indicates the coefficient is different from 0 with at least 99 percent confidence.

Distance

Specification 1 allows us to investigate how transportation costs per pound vary with flight time. The elasticity estimates range between 0.30 and 0.42, suggesting that increasing flight time by 10 percent should increase TEP transport costs by 3 to 4 percent, with our most reliable estimates falling at the lower end of this range.

Quantity Effects

Specifications 1 and 2 indicate that increasing the quantity of cargo put out to bid through TEP by 10 percent should decrease the cost per pound transported by 3 percent. Consequently, combining cargo shipments that would otherwise be placed days apart for the same origin-destination pair can generate cost savings. This, of course, comes at the expense of delayed cargo arrival times and therefore represents a trade-off. We have no evidence that this trade-off is not currently being balanced appropriately.

Number of Bidders

The number of bidders in the tender has a significant effect on the price that must be paid. The point elasticity pertaining to the nmber of bidders varies between –0.38 and –0.55, depending on the specification and covariates included, but our most reliable estimates fall at the lower end of this range.

When more carriers bid on a tender, costs can decline for two reasons. First, if each carrier's cost on a tender is stochastic and if carriers bid based on their actual cost, having more carriers bid gives the government a greater chance of receiving low-cost bids. Second, as the number of bidders increases in general on a route, competitive pressures to bid in line with actual costs may also increase, causing carriers to bid closer to their true costs.

The estimates in Table B.3 clearly suggest that increasing the number of bidders on tenders will have positive effects on program cost. To understand how demand for tender services affects the number of carriers who bid on a tender, we ran models of the form

$$\log(N_{odt}) = \rho + \vartheta \log(Q_{od}) + \sigma_{CO} + \psi_{CD} + \varepsilon_{odt},$$

where Q_{od} is the total number of tenders between origin-destination *od* from October 1, 2008, through December 17, 2009, and σ_{CO} and ψ_{CD} are fixed effects for the country in which the origin and destination airports, respectively lie, and ε_{odt} is a random error term. We estimated the model using ordinary least squares. We also estimated a second version of the model using Poisson regression that respects the fact that N_{odt} only takes integer values. The coefficient ϑ (i.e., the coefficient on $\log(Q_{od})$) can be interpreted as an elasticity in both the ordinary least squares and the Poisson models. We also estimated the model with and without the country origin and destination fixed effects. Table B.3 shows the results of these regressions.

The estimates in Table B.3 suggest that doubling the number of tenders between an *od* pair would be expected to increase the number of bidders on that route by 13 to 19 percent on average, with our most reliable estimates falling at the lower end of this range.

Some caution is warranted in interpreting this result. First, TEP currently has only seven qualified carriers, so there is clearly an upper bound on the number of carriers that can currently bid on tenders. One strategy for reducing costs might involve expanding the program beyond seven qualified carriers. There is, of course, a limit to the benefits that increasing the number of carriers that bid on tenders can offer; however, the statistical analysis does not suggest that that limit has yet been reached.

Table B.3
Regression to Predict Number of Bidders

Model	1a (OLS)	1b (OLS)	2a (Poisson)	2b (Poisson)
Dependent variable	$\log(N_{odt})$		N_{odt}	
υ (Coef. on $\log(Q_{od})$)	0.192**	0.149**	0.158**	0.1347**
Standard error	0.004	0.005	0.004	0.0049
Origin and destination dummies	No	Yes	No	Yes
Observations	17,725	17,725	17,725	17,725
Parameters	2	22	2	22
R^2	0.123	0.232	NA	NA
Log-likelihood	NA	NA	–35,771.1	–34,957.4

NOTE: Within this table, * indicates the coefficient is different from 0 with at least 95 percent confidence, and ** indicates the coefficient is different from 0 with at least 99 percent confidence.

Analysis of Tender Delivery Times

A review of the tender data suggests that TEP generally required cargo to be delivered within 96 hours. If delivery takes longer than 96 hours, the carrier has breached its tender agreement, unless it has been provided an exemption. Exemptions can be granted for a variety of reasons, including weather, runway closures, airport construction, or loading and unloading issues.

Table B.4 indicates that less than 60 percent of all tenders are delivered within the 96-hour window. Approximately 33.5 percent of all tenders are late and nonexempt, while 9 percent of all tenders are late but granted an exemption.

The mean delivery time for TEP is 87.6 hours, but the standard deviation is quite large, at 87.7 hours. Figure B.5 shows the distribution of delivery times under TEP. The data used to construct this graph were restricted to the May 1, 2009, to December 17, 2009, period because data from earlier periods did not report the number of days a delivery was late for deliveries that took longer than 96 hours.

A number of factors could contribute to poor delivery time performance. Delivery time may be related to the workload a winning carrier faces. Furthermore, conditions outside the control of the carrier and that lead to delivery time exemptions are another example of potential causes of longer delivery times. Unfortunately, the tender data provide limited insight into what caused poor delivery time performance. To try to understand the drivers of poor performance, we ran regressions in which the dependent variable was either an indicator variable of whether the cargo delivery was late or not (i.e., delivery time exceeds 96 hours) or the natural log of the delivery time. In the latter case, we had to restrict our attention to the May 1, 2009, to December 17, 2009, period because the data on delivery time for earlier periods was censored at 96 hours. Our explanatory variables included

- a variable intended to measure workload, which equals the natural log of the number of tenders the carrier won over the previous and following two-day period

Table B.4
Share of Tenders That Meet or Breach Delivery Time Requirements (October 1, 2008, to December 17, 2009)

	Nonexempt		Exempt	
	Number	Percent	Number	Percent
Delivery time meets requirement	9,874	55.7	314	1.8
Delivery time breaches requirement	5,930	33.5	1,609	9.1

Figure B.5
Distribution of Tender Delivery Times (May 1, 2009, to December 17, 2009)

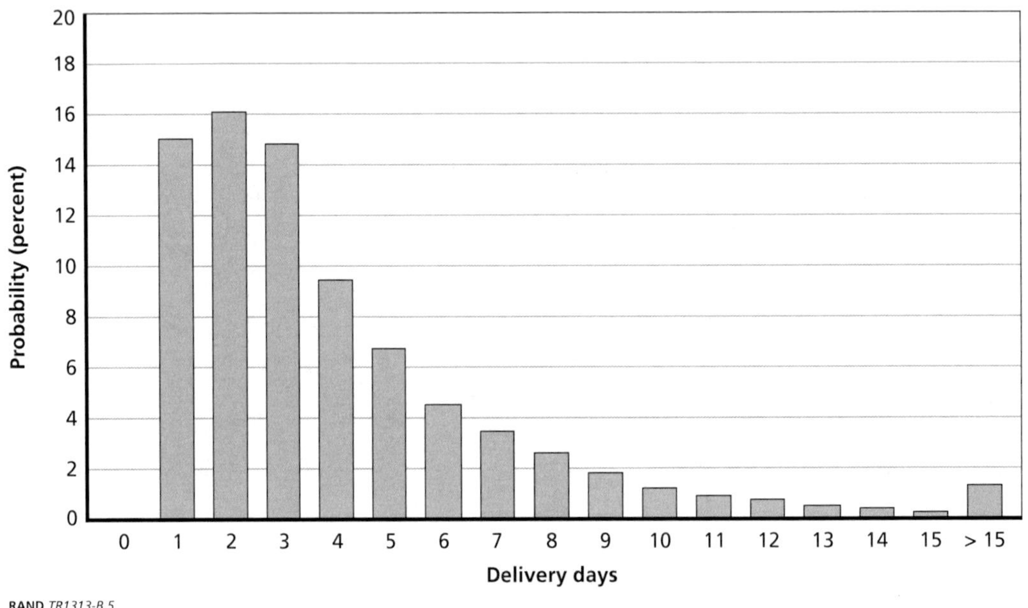

RAND *TR1313-B.5*

- a dummy variable indicating whether the delivery was exempt from meeting its delivery time requirement
- carrier-specific dummy variables to capture general carrier performance differences
- airport origin and destination dummy variables to capture factors that may be specific to airports that affect delivery time.

The results of the regression analysis are presented in Table B.5. These results suggest the following:

- The coefficient on the workload variable is not statistically different from 0 in either of the regressions, suggesting that carrier workload does not contribute directly to poor (or good) carrier performance measured in terms of delivery times.
- If a delivery receives a delivery time exemption, it increases the probability that the shipment will be late by 40 percent. The second regression suggests that tenders that receive a delivery time exemption are likely to take 88 percent longer to arrive. Because exemptions are presumably given for legitimate reasons, this finding has few policy implications.

Table B.5
Regression to Predict Late Deliveries

Dependent Variable	Dummy Variable = 1 if Late, 0 Otherwise	Log (delivery time)
Log (number of tenders within 2 days pre and post)	0.008	−0.010
Standard error	0.007	0.022
Exemption dummy variable	0.406**	0.879**
Standard error	0.011	0.033
Carrier dummies (excluded carrier = G)		
Carrier A	−0.084**	−0.178**
Standard error	0.014	0.040
Carrier B	0.107**	0.280**
Standard error	0.015	0.046
Carrier C	0.016	-0.104
Standard error	0.016	0.066
Carrier D	−0.049**	−0.248**
Standard error	0.013	0.042
Carrier E	−0.226**	−0.466**
Standard error	0.011	0.034
Carrier F	0.013	0.036
Standard error	0.012	0.036
Origin and destination dummies	Yes	Yes
Observations	17654	9671
Parameters	64	62
R^2	0.193	0.242

NOTE: Within this table, * indicates the coefficient is different from 0 with at least 95 percent confidence, and ** indicates the coefficient is different from 0 with at least 99 percent confidence.

- After controlling for workload, exemptions, and origin and destination factors that may influence delivery times, some carriers appear to be performing better than others. In particular, carrier E is performing the best, while carrier B is performing the worst, in terms of delivery time performance. For instance, the first regression shown in Table B.5 suggests that the likelihood that carrier B is late is 33 percent (0.107 + 0.226) greater than the likelihood that carrier E would be late.

APPENDIX C

Estimating the Full Marginal Costs of Utilizing C-130 Aircraft

This analysis provides estimates of the full marginal cost of changing the rate at which C-130 aircraft fly when home stationed or deployed. Obviously, operating an aircraft at a higher operating tempo will incur additional direct expenditures related to the additional flying. The additional flying would also hasten the time until major maintenance and aircraft replacement, which will induce additional expenditures.

To study how a marginal increase in FH affects the full range of costs the Air Force incurs, we developed a discounted cash flow model for conducting the airlift activities the deployable fleet of C-130E and C-130H aircraft currently perform. We used the model to estimate the full marginal cost of an additional one-time flight. The cost estimates presented here can inform decisionmakers wishing to compare the cost of utilizing C-130 aircraft relative to other means of transport.

The data used in this analysis come from a variety of sources, including

- USAF C-130E and C-130H aircraft inventory and projections of future aircraft life and use
- USAF tabulations of home stationed and deployed FH intensities and SF by squadron
- cost estimates developed from the CORE model
- reports from RAND and other defense research organizations that provide estimates of major maintenance costs and timing.

Our analysis was confined to a subset of the C-130 fleet. First, the analysis considered the current fleet of MDS C-130E and C-130H aircraft. Data on the current fleet of C-130J aircraft are maintained separately and were not available for our analysis. Second, since our focus was on C-130 use for typical cargo movements, we did not consider any of the modified variations on the C-130E and C-130H aircraft, such as the AC-130 or WC-130. Third, we limited our attention to aircraft associated with commands that could potentially be deployed to combat areas—AFRC, AMC, ANG, PACAF, and USAFE. We excluded aircraft associated with Air Combat Command, Air Education and Training Command, Air Force Materiel Command (AFMC), and Air Force Special Operations Command.

The appendix proceeds by describing the current fleet of C-130E and C-130H aircraft and the assumptions about future aircraft utilization we received from the Air Force. Next, we describe the cost assumptions used in the analysis, taking considerable time to describe how we estimated fixed and variable costs for aircraft associated with different commands and whether these aircraft are deployed or home stationed. Next, we describe the discounted cash flow model we used to generate estimates of the marginal cost of different aircraft utilization assumptions. Finally, we summarize the marginal cost estimates we obtained using the discounted cash flow model.

The Fleet of C-130E and C-130H Aircraft

The inventory of C-130Es and C-130Hs considered for this analysis consists of 317 aircraft. AMC created the corresponding aircraft data set on June 17, 2009, and provided it to RAND. This data set offers a snapshot of what the fleet of aircraft looked like as of that date. Each aircraft is associated with a command, owning unit, and possessing unit in the inventory. For each aircraft, the data include EBH, which accumulates according to its past FH and the SF associated with past flights. In particular, a single flight will contribute to an aircraft's EBH in an amount equal to the SF associated with that flight times the duration of the flight in hours. The inventory provides information on the lifetime and recent average SF each aircraft has experienced, along with the aircraft's squadron-average SF.

In addition to providing information on the historical use of the aircraft, the inventory provides an estimate of each aircraft's FH and years until grounding. The Air Force assumes grounding occurs once an aircraft reaches 45,000 EBH. For our analysis, we used the Air Force's assumptions about future aircraft utilization. In particular, we calculated an aircraft's future FH/year and average FH SF as follows and incorporated it into our life-cycle cost analysis:

$$\text{Future average FH/year} = \frac{\text{Est FH until grounding}}{\text{Est years until grounding}}$$

$$\text{Future average SF} = \frac{45{,}000 - \text{Current EBH}}{\text{Est FH until grounding}}$$

Table C.1 summarizes the aircraft counts, EBH, and future FH and SF we inferred from the Air Force aircraft inventory. Two hundred and sixty-three of the 317 aircraft in the inventory are of the newer C-130H aircraft design. Aircraft were not distributed evenly across the commands we considered. ANG operates the most aircraft, with 141 C-130Es and C-130Hs; PACAF and USAFE operate the least aircraft, with 14 and 13 aircraft, respectively. All the aircraft PACAF and AFRC operate are C-130Hs, while USAFE operates only C-130Es.

Aircraft in the current C-130 fleet date back to as early as the 1960s. As a result, many C-130s, especially C-130Es, have been in use for decades. In general, C-130Es tend to have more EBH than C-130Hs, although there is considerable variation in EBH across and within commands for each MDS. Figure C.1 shows the cumulative distribution of EBH by command. ANG and AFRC both operate aircraft that tend to have fewer EBH than those operated by the other commands considered in this analysis.

Aircraft associated with AMC and PACAF are assumed to perform the most FH per year, while ANG is projected to fly the fewest FH per aircraft per year. In terms of FH SF, flights out of AMC are projected to be the least detrimental to aircraft life, while ANG flights will reduce aircraft life at a faster rate per FH than other commands.

Table C.1
Summary of C-130E and C-130H Aircraft Inventory and Future Use Assumptions

			Future Average	
	Number of Aircraft	Average EBH per Aircraft	FH per Year per Aircraft	SF per FH
C-130E				
AFRC	0	NA	NA	NA
AMC	23	34,688	664.7	1.8
ANG	18	32,835	419.5	2.1
PACAF	0	NA	NA	NA
USAFE	13	37,271	578.1	2.2
All C-130Es	54	34,692	562.1	2.0
C-130H				
AFRC	84	17,748	435.9	2.2
AMC	42	35,954	767.5	2.3
ANG	123	17,825	379.5	2.6
PACAF	14	39,100	729.6	2.2
USAFE	0	NA	NA	NA
All C-130Hs	263	21,828	478.3	2.4
C-130E and C-130H combined				
AFRC	84	17,748	435.9	2.2
AMC	65	35,506	731.1	2.1
ANG	141	19,741	384.6	2.5
PACAF	14	39,100	729.6	2.2
USAFE	13	37,271	578.1	2.2
All C-130Es and C-130Hs	317	24,019	492.4	2.3

SOURCE: All values derived from USAF C-130 aircraft inventory data maintained by AMC. EBH calculated as of June 17, 2009.

Differences in Deployed versus Home Stationed Flying

Flying requirements are known to vary when aircraft are deployed as opposed to at home station. To understand these differences, we used data from AFMC on average squadron FH per day and SF when home stationed and deployed. The data are organized by squadron and cover only C-130H aircraft.[1] As a result, no data for USAFE were available because its entire fleet of C-130 aircraft consists of the "E" series.

Table C.2 summarizes the average FH per day and SF calculated from the AFMC data. Aircraft from every command tend to fly more when deployed than when at home station. The biggest increase over home station occurred in AFRC, while PACAF had the smallest increase. In terms of SF, flights for deployed AFRC aircraft were less intense than those for home stationed aircraft. The data suggest that the opposite holds true for aircraft associated with the AMC, ANG, and PACAF.

[1] We were unable to identify a source that provides these data for C-130Es.

Figure C.1
Distribution of C-130E and C-130H Equivalent Flying Hours Across Commands, as of June 17, 2009

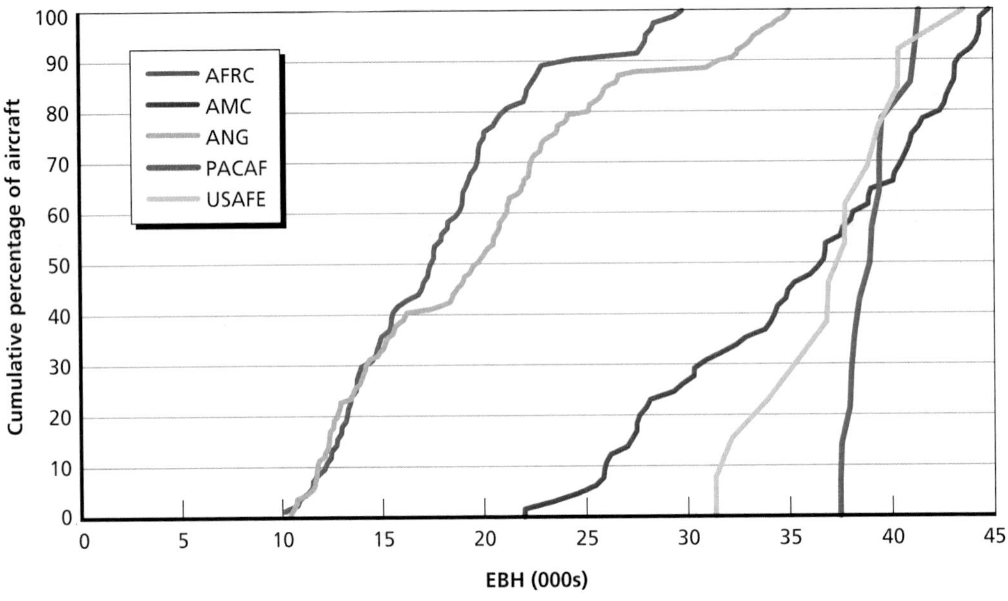

RAND *TR1313-C.1*

Table C.2
Summary of Flying Hour Intensities and Severity Factors for Home-Stationed and Deployed C-130Hs

	AFRC	AMC	ANG	PACAF
Number of squadrons with data	9.0	3.0	16.0	1.0
FH/day home stationed	1.4	1.7	1.4	2.2
FH/day deployed	5.1	3.7	4.2	4.1
Ratio of deployed to home stationed	3.7	2.2	3.0	1.8
SF home stationed	2.5	2.6	2.5	2.2
SF deployed	1.6	3.2	3.0	2.7
Ratio of deployed to home stationed	0.6	1.2	1.2	1.2

NOTE: Based on calculations performed for C-130H aircraft associated with different squadrons. Data from AFMC, August 10, 2010.

To understand the level of deployment in the Air Force's assumptions for future flying activity, we compared the historical C-130H FH (Table C.2) with the Air Force assumptions about future C-130H annual FH (Table C.1), by command. For AFRC, ANG, and PACAF, the Air Force has assumed future FH will be lower, on average, than the home-stationed rate of the data presented in Table C.2. For consistency, in our base case scenario, we made the assumption that C-130s would be deployed 0 percent of the time in the Air Force's projections of future FH. To identify the marginal cost of deploying aircraft in the future, we increased our assumption of the percentage of time aircraft are deployed from that 0 percent.

Cost of C-130 Operations

This section describes the costs associated with operating and maintaining a fleet of C-130 aircraft. These estimates, along with the future flying projections in the previous section, drive the discounted cash flow model. While our aircraft inventory does not include any information on the Air Force's C-130J fleet, this section also includes cost estimates for this aircraft series. We needed these estimates to project the cost of maintaining C-130 airlift capabilities after a C-130E or C-130H must be grounded.

Our analysis distinguished between fixed and variable aircraft operating costs to help us understand how these costs might vary by command when an aircraft is at home station or deployed. We also estimated the effects of additional FH on major maintenance costs and timing, as well as on aircraft replacement costs. This section describes each of these cost categories and the discount rate assumption for this analysis.

Operating Costs

To estimate operating costs, we modified a standard USAF model called CORE. Our analysis broke the resulting cost estimates out into the fixed and variable costs. The variable costs are those associated with additional flying (i.e., those that would change if the cargo mission were added to an already deployed force), while the fixed costs are those for maintaining the aircraft at either the home station or a deployed location. Because personnel cost factors change when a unit is deployed (as explained later), the fixed costs at a deployed location may differ from those at home station.

The CORE model uses standard USAF planning factors published in AFI 65-503 appendixes to estimate the marginal cost of adding a squadron to (or removing it from) an existing active-component base. In essence, the model uses resource counts, resource consumption rates, and resource prices to estimate the costs in a format defined by the Air Force Cost Analysis Improvement Group (AFCAIG). The costs are grouped into seven distinct categories with multiple subcategories. The major AFCAIG categories are

- 1.0 Personnel (direct compensation, including health and pensions)
- 2.0 Materiel support (materiel consumed at the squadron)
- 3.0 Intermediate maintenance (maintenance and materiel consumed outside the squadron)
- 4.0 Depot maintenance (aircraft and engine overhaul, mainly)
- 5.0 Contractor support (both depot and intermediate maintenance, but performed at contractor locations)
- 6.0 Sustainment (mainly support equipment replacement, sustainment engineering, and software)
- 7.0 Personnel support (training, permanent change of station, medical, base operating support [BOS], and installation support).

As an example of the subcategories, the first indenture of the personnel category includes subcategories for operations, base maintenance, squadron staff, and security personnel. BOS personnel costs are included in the personnel support category. There are even subsubcategories. For example, the operations subcategory has subsubcategories that distinguish among pilots, other officer aircrew members, and enlisted aircrew members.

There are specific prices for each resource category, often tied to the specific aircraft MDS being evaluated. Thus, pilots may receive the same remuneration as other officers, but their training costs are typically much higher, and the cost of training a fighter pilot typically exceeds the costs of training a transport aircraft pilot.

The standard CORE model was developed expressly to model active component squadrons' costs. Over time, the reserve component (RC) has begun to comprise a larger portion of the total force. The RC possesses and operates the majority of C-130 aircraft and has supported a major portion of recent contingency operations. We have expanded some of the resource categories in CORE and used pricing data from the RC appendixes in AFI 65-503 to obtain a more-accurate picture of the ANG and AFRC costs than would be possible by relying on active component personnel costs.

First, we expanded the personnel definitions in CORE to include drill personnel whose peacetime activities are limited to periodic weekend and summer training activities. We then revised the CORE calculations to include the cost of drill personnel in each of the personnel categories.

While the standard ANG and AFRC tables in AFI 65-503 identify the numbers of drill, Active Guard Reserve (AGR), and Air Reserve Technician (ART) personnel in typical units at their peacetime locations, they do not indicate how many drill personnel change to active status when they deploy. Drill personnel changing to either ART or AGR status receive that level of compensation. Thus, we modified the CORE calculations in two ways, one for operations personnel and another for maintenance personnel. For operations personnel, we used the internal CORE calculations based on crew size, crew ratio, and number of aircraft to compute the number of ART or AGR personnel needed to deploy a squadron. Then we increased the AGR or ART personnel by that number and reduced the number of drill person by the same amount. For maintenance personnel, we used the START model to estimate the additional number of maintenance personnel needed at a typical deployed unit and adjusted the ART, AGR, and drill maintenance personnel accordingly.

In addition, CORE's focus on squadron costs uses the primary aircraft authorized (PAA) concept for estimating resources and costs. PAA is only a subset of the total aircraft inventory (TAI) owned by the Air Force. The PAA concept is used for budget planning and other unit-level forecasts because some additional aircraft assigned to the unit may be assigned to other activities, such as depot maintenance.

Assumptions About Some CORE Personnel Factors

While the model is formally defined in one of the AFI 65-503 appendixes (Appendix A56-1), the other AFI 65-503 appendixes that provide the requisite data have progressively deviated from that formal standard over time. More-recent appendixes often do not provide sufficient detail for entry directly in the model, especially about personnel levels. In perhaps the most egregious example of this steady erosion of detailed information, the active component appendix that identifies typical aircraft unit strengths (A42-1) no longer presents any data regarding squadron personnel levels at all.

It was necessary to make some assumptions about the available data and to use older data to estimate some C-130 parameters in CORE. In the specific case of active component personnel data, we used the personnel data from the version of A42-1 originally published in January 2005, implicitly assuming that there had been no meaningful changes in actual squadron manning assignments since then.

Both the ANG and the AFRC provide more-recent personnel data but not at the level of detail needed for CORE. ANG provides an aggregate weapon system manpower category with no separation of personnel into CORE's operations, maintenance, staff, and security subcategories, along with a BOS category. Therefore, we assigned all weapon system manpower AGRs to full-time staff positions (e.g., wing commander or maintenance director with administrative personnel), and we assigned sufficient drill personnel to meet the crew ratio to operations. Remaining personnel, including all civilian employees, were assigned to maintenance.

AFRC provides more detailed data about operations and maintenance personnel, distinguishing between operations and maintenance. The command also has an additional category, support and wing staff, that subsumes the staff positions, BOS, and security. We assigned all such personnel to wing staff, essentially assuming that none of them would deploy during a contingency and that they would still conduct their activities at home station.

In addition, the AFRC appendix makes no distinction between ARTs who are officers and those who are enlisted in each of the subcategories. We used the ratio of drill officers and enlisted personnel in each subcategory to assign ARTs to either officer or enlisted.

Assumptions About Personnel Deployment

Personnel costs for ANG and AFRC units change during a contingency because some deploying personnel must be changed to AGR or ART status. As mentioned above, we used the START model to estimate the number of personnel required for a typical deployment. Specifically, we assumed that 16 aircraft would deploy. Because typical ANG or AFRC squadrons used in our CORE model have only eight PAA, we assumed that these squadrons would provide half the operations and maintenance personnel for such a deployment and that they would join an existing base with adequate wing staff, security, and BOS. Thus, these squadrons only needed to activate operations and maintenance personnel.

To estimate the costs of activating these personnel in the CORE model, we shifted sufficient operations drill personnel to ART or AGR status to meet the same crew ratio as peacetime. For ANG units, this was the entire population of operations personnel. Some AFRC operations personnel did not deploy but continued their drill and administrative activities at home station. We used the START model's maintenance requirement to estimate the number of maintenance personnel that would change from drill to ART or AGR status.

Assumptions About Training

Many, but not all, RC personnel have had basic and initial skill training during their prior service. Because CORE was originally designed for the active component, it has no factor to estimate the costs of training non–prior service drill personnel in their initial skill. We assumed that all RC personnel had prior service for the purposes of this analysis. Thus, we assumed that the ANG and AFRC had zero training costs. To the extent that each MAJCOM requires non–prior service personnel, it will experience higher fixed costs than estimated in this analysis.

Implicitly, we also assumed that the training costs do not change regardless of the number of aircraft deployed. That is, we assumed that the personnel annual attrition rates and the need for replacement personnel would remain constant, so Air Education and Training Command costs would not change, which means that the AFCAIG training category costs would remain the same in the active component.

Distinguishing Between Fixed and Variable Costs

The AFCAIG categories and subcategories provide a natural division of fixed and variable costs. Four of the categories (personnel, intermediate maintenance, sustainment, and personnel support) are fixed costs associated with owning and operating the fleet. One of the categories (materiel support) consists solely of costs that rise with increasing FH. The two remaining categories (depot maintenance and contractor support) are mainly composed of subcategories that do not change with FH, but each has an exception. In the case of depot maintenance, there is an accounting category for FH-dependent costs, which consists mostly of engine maintenance. For contractor support, the contractor logistics support (FH) subcategory reflects support contract provisions that are priced by FH (often related to component maintenance and materiel replacement).

Estimates of Fixed and Variable Costs

Table C.3 summarizes the estimates of fixed costs when the aircraft is home stationed or deployed, as well as the variable costs per flying hour. The fixed costs are annualized for incorporation into the discounted cash flow model presented shortly.

In most cost categories, the C-130H aircraft is slightly more expensive than the older C-130E model. The C-130J aircraft, however, generally provides a significant cost advantage over the C-130E and C-130H aircraft, in terms of both fixed and variable costs.

Table C.3
Estimates of C-130 Fixed and Variable Costs

	Fixed Cost per Year ($000s, FY 2010)		Variable Cost per Flying Hour ($000s, FY 2010)
	At Home Station	Deployed	
C-130E			
AFRC	4,924.9	6,716.0	5.8
AMC	5,153.4	5,153.4	5.8
ANG	5,056.9	7,522.9	5.8
PACAF	4,072.9	4,072.9	5.8
USAFE	5,439.8	5,439.8	5.8
C-130H			
AFRC	4,945.1	6,414.5	5.9
AMC	5,172.1	5,172.1	5.9
ANG	5,077.8	8,243.8	5.9
PACAF	4,091.6	4,091.6	5.9
USAFE	5,458.5	5,458.5	5.9
C-130J			
AFRC	4,593.0	6,289.1	3.3
AMC	3,766.3	3,766.3	3.3
ANG	4,491.7	6,941.5	3.3
PACAF	3,380.6	3,380.6	3.3
USAFE	4,687.1	4,687.1	3.3

NOTE: Estimates developed using CORE model and assumptions described previously.

The cost of the C-130J compared to the C-130E and C-130H at AMC is worth highlighting. For example, at AMC, the C-130H generates $5.2 million per year in fixed costs annually (whether deployed or at home) and produces variable costs at a rate of approximately $5,900 per FH. If AMC replaces a C-130H with a C-130J, the fixed costs associated with operating the aircraft drop to $3.8 million per year, and variable costs fall to $3,300 per FH. Recognizing this cost savings and its particularly large effect at AMC is helpful for interpreting some of the results presented at the end of this appendix.

Major Maintenance Costs

Corrosion and fatigue from age and use damage aircraft, leading to flight restrictions, major maintenance activities, and ultimate replacement. In particular, the C-130's center wing is susceptible to significant damage from use. For a detailed discussion of the corrosion and fatigue issues the fleet of C-130 aircraft faces, see Orletsky et al., 2011. These issues have caused the Air Force to adopt the following major maintenance schedule to address fatigue issues that occur with the C-130's center wing:

- Rainbow Fitting Replacement: The outer-wing attachment points (known as *rainbow fittings*) do not have a fatigue life commensurate with the rest of the center wing and must be replaced at about 24,000 EBH.
- TCTO 1908 Inspection and Repair: The Air Force applies TCTO 1908 protocol to address fatigue risks present at 38,000 EBH. Assuming the aircraft passes the TCTO 1908 protocol, it is allowed to operate unrestricted to the assessed service-life limit of the center wing, 45,000 EBH.

We assumed that all aircraft would undergo rainbow fitting replacement at 24,000 EBH and successfully complete TCTO 1908 at 38,000 EBH, allowing them to fly to 45,000 EBH unrestricted. Furthermore, we assumed that C-130J aircraft would need to undergo similar major maintenance at the same costs and EBH points as earlier C-130 variants. Table C.4 provides the cost and timing of both major maintenance activities.

At 45,000 the C-130 center wing must be replaced under current component design limitations. The cost of the center wing replacement is estimated at $9.4 million in FY 2010 dol-

Table C.4
Timing and Cost of Standard Major Maintenance Activities for C-130 Aircraft

Maintenance Activity	Timing (EBH)	Cost ($000s, FY 2010)
Rainbow fitting	24,000	733.8
TCTO 1908/Lower wing	38,000	932.9

SOURCE: Based on estimates in Orletsky et al., 2011. Costs inflated from FY 2007 dollars to FY 2010 dollars using standard Air Force cost inflation factors for operations and maintenance activities.

lars.[2] In addition to costing millions of dollars, this replacement can take a significant amount of time. Furthermore, even after the center wing is replaced, the aircraft is likely to be at greater failure risk per FH. Because center wing replacement is not currently standard practice and because our model is not well suited for modeling the costs associated with increased failure risk with EBH, we did not consider replacement of the center wing. Should the center wing replacement occur, the next major scheduled maintenance is replacement of the outer wing, which occurs at 60,000 EBH under current component design limitations.

Aircraft Replacement Costs

Current protocol calls for replacing C-130 aircraft once they reach 45,000 EBH. The current replacement version of the C-130 aircraft is the C-130J. In our analysis, we assumed that all C-130E and C-130H aircraft would be replaced with C-130J aircraft at 45,000 EBH. Furthermore, because we looked into perpetuity, we had to make assumptions about C-130J replacements in the distant future. For the purposes of our analysis, we assumed that C-130Js would be replaced with C-130Js of the same cost (in real dollars) at 45,000 EBH.

Orletsky et al., 2011, provides an estimate for the flyaway cost of a C-130J of $61 million in FY 2007 dollars. We inflated this estimate to FY 2010 dollars using standard USAF inflation factors for aircraft and missile procurement, to obtain an estimate of $63.9 million in FY 2010 dollars.

Discount Rate

In this analysis, we considered a variety of different costs that will occur at different times in the future. To put these costs in comparable present dollar terms, we used the long-term real discount rate of 2.7 percent per year that the Office of Management and Budget prescribes (OMB, 2009). OMB based its recommended real discount rate for long-term investments on the 30-year Treasury bill rate and assumptions about inflation. This rate is appropriate for cash flows that are known with certainty or when decisionmakers are not risk adverse. If there is uncertainty about the future cash flows under consideration and when decisionmakers are risk averse, a higher discount rate can be justified. For the purposes of sensitivity analysis, we also considered a 5-percent real discount rate.

Modeling C-130 Cost from a Discounted Cash Flow Perspective

This section describes the discounted cash flow model we applied in our analysis. In the model, time is continuous and indexed by t. One unit of time represents one year in the model. The Air Force operates a fleet of N similar aircraft. Let $n = 1,2,...N$ be an index of C-130E and C-130H aircraft in the fleet. We assumed the fleet to be operating in a steady-state environment with annual FH and average SF remaining unchanged after $t > 0$. In particular, we assumed that the assumptions of future aircraft use provided by the Air Force hold into perpetuity.

Each aircraft in the fleet is associated with EBH, $E_n(t)$, which increases over an aircraft's life with usage. We started the calculations at $t = 0$, which we assumed to be the end of FY 2010. We took the set

[2] Orletsky et al., 2011, estimates a cost of $9 million dollars for a center wing replacement. We have updated that cost estimate to FY 2010 dollars.

$$\{E_1^0, E_2^0, \ldots., E_N^0\}$$

as our initial condition, with

$$E_n(0) = E_n^0$$

for all n. To calculate the set

$$\{E_1^0, E_2^0, \ldots., E_N^0\}$$

as of the end of FY 2010, we updated the EBH of each aircraft as of June 17, 2009, using the assumptions about future FH rates and SFs.

There is a fixed cost associated with the fleet that accrues at rate f_n per aircraft over time. The fixed cost f_n varies by aircraft, depending on its command and the percentage of time it is assumed to be deployed or stationed at home. In particular, if ϕ_n represents the percent of time aircraft n is deployed, then

$$f_n = \phi_n f_n^D + (1 - \phi_n) f_n^H,$$

where

f_n^D = the fixed cost per year when deployed

f_n^H = the fixed cost per year when stationed at home.

Each aircraft n engages in FH at a particular rate h_n, of a particular severity factor s_n, and of a cost that varies proportionally with FH by c_n (i.e., 1 FH costs, c_n). The increase in EBH for aircraft n of 1 hour of flying is equal to s_n. For our analysis, we assumed that the Air Force estimates of future C-130 flight activity did not reflect deployed activities (i.e., $\phi_n = 0$ in our base case). A simplistic comparison of C-130 deployed and at-home FH with the assumptions of future aircraft FH supports this assumption. For the purposes of varying the deployment rate, ϕ_n, we calculated

$$h_n = \phi_n \mu_n h_n^H + (1 - \phi_n) h_n^H$$

and

$$s_n = \phi_n \omega_n s_n^H + (1 - \phi_n) s_n^H,$$

where

h_n^H = the FH rate of aircraft n when stationed at home

s_n^H = the severity factor of aircraft n when stationed at home

and μ_n and ω_n are the ratio of deployed to home-based FH and SF described in Table C.2 for the command associated with aircraft n.

Major maintenance events occur at certain points in an aircraft's life, depending on an aircraft's EBH. Let $i = 1, 2, \ldots, I$ be an index on the set of major maintenance events, and let α_i be the EBH that triggers major maintenance event i. Assume that α_i is ordered such that $0 < \alpha_i \leq \alpha_2 \leq \ldots \leq \alpha_I$. Let the cost of major maintenance event i equal θ_i. For the purposes of our analysis, we considered only the major maintenance events in Table C.4.

We assumed that, once an aircraft reaches EBH of b, it must be replaced. As discussed above, b = 45,000 EBH. The capital cost of replacing the aircraft is p. Furthermore, we assumed that, after replacement, the new aircraft would be operated at the same FH intensity (h_n) and under the same SF (s_n). We assumed that the replacement aircraft would be C-130Js, which may operate at different costs but are maintained and replaced under the same EBH schedule. We represented C-130Js in the calculations with a superscript, J.

Finally, to discount future costs to current dollars we assumed a real discount rate, r. Table C.5 provides a list of the key variables integrated into the analysis and their sources.

Table C.5
Summary of Data Requirements and Sources for Discounted Cash Flow Model

Variable	Description	Source
$\{E_1^0, E_2^0, \ldots, E_N^0\}$	EBH for each aircraft in current fleet	USAF C-130 TAI spreadsheet updated to the end of FY 2010 using assumptions on future flying hour intensities and SF
$\{h_1^A, h_2^A, \ldots, h_N^A\}$	Forecast future flying hour intensities by aircraft when home stationed	USAF C-130 TAI spreadsheet
$\{\mu_1, \mu_2, \ldots, \mu_N\}$	Ratio of deployed FH/day to home stationed FH/day	USAF at home and deployed C-130 spreadsheet
$\{s_1^A, s_2^A, \ldots, s_N^A\}$	Forecast of future average SF by aircraft when home stationed	USAF C-130 TAI spreadsheet
$\{\omega_1, \omega_2, \ldots, \omega_N\}$	Ratio of deployed SF to home-stationed SF per flight hour	USAF at home and deployed C-130 spreadsheet
$\{c_1, c_2, \ldots, c_N\}$	Variable cost of a flying hour by aircraft	Modified CORE Model
$\{f_1^H, f_2^H, \ldots, f_N^H\}$	Fixed cost per year by aircraft home stationed	Modified CORE Model
$\{f_1^D, f_2^D, \ldots, f_N^D\}$	Fixed cost per year by aircraft deployed	Modified CORE Model
$\{\phi_1, \phi_2, \ldots, \phi_N\}$	Percent of time aircraft are deployed	Authors based on assessment of home based and deployed FH relative to Air Force's assumptions of future FH
$\{\alpha_1, \alpha_2, \ldots, \alpha_I\}$	Set of EBH that trigger major maintenance activities	Orletsky et al., 2011
$\{\theta_1, \theta_2, \ldots, \theta_I\}$	Cost of major maintenance type	Orletsky et al., 2011, updated to $ FY 2010
β	EBH at which an aircraft must be replaced	Orletsky et al., 2011
π	Capital cost of replacing an aircraft	Orletsky et al., 2011, updated to $ FY 2010
r	Real discount rate	OMB

Calculating Discounted Future Cash Flows

For calculating discounted cash flows, it was helpful to track the *nth* aircraft's vintage. Let $\nu_n(t) = 0,1,2,\ldots$ equal the number of times the *nth* aircraft has been replaced as of time t. To simplify notation, assume that $\nu_n(0) = 0$ for all n. Under this assumption, the EBH of the *nth* aircraft at any time is

$$E_n(t) = E_n^0 + h_n s_n t - v_n(t)\beta.$$

Note that an aircraft's life span (measured in time) is given by

$$L_n = \frac{\beta}{h_n s_n}.$$

Aircraft n is initially replaced at time

$$T_n = \frac{\beta - E_n^0}{h_n s_n}.$$

Therefore, $\nu_n(t)$ is given by the function

$$v_n(t) = \left[v \in \{0,1,\ldots\} : T_n + (v-1)L_n \le t < T_n + vL_n\right].$$

Major maintenance event i is triggered when $E_n(t) = \alpha_i$. For vintage 0, the *ith* major maintenance event occurs at time

$$A_{in} = \frac{\alpha_i - E_n^0}{h_n s_n},$$

which may be before $t = 0$. To identify which major maintenance is scheduled to occur first after $t = 0$, we solved for

$$i_n^0 = \left[i = 1,2,\ldots,I : A_{(i-1)n} \le 0 < A_{in}\right],$$

where $A_{0n} = T_n - L_n < 0$ is added to the set $\{A_{1n}, A_{2n}, \ldots, A_{in}\}$ and represents the date at which the initial vintage of aircraft n was purchased.

We were then able to characterize the life-cycle costs of an aircraft, taking into account replacement and continuation costs. It is convenient to write the life-cycle cost in recursive form. Specifically,

$$\Pi_n(E_n^0) = \overbrace{\int_0^{T_n} e^{-rt}\left[h_n c_n + f_n\right] t\, dt}^{\text{PDV of O\&M cost on Initial Aircraft}} + \overbrace{\sum\nolimits_{i=i_n^0,\ldots,I} e^{-rA_{in}}\theta_i}^{\text{PDV of Major Maintenance Cost on Initial Aircraft}} + \overbrace{e^{-rT_n}\pi}^{\text{PDV of Replacement Cost for Initial Aircraft}} + \overbrace{e^{-rT_n}\Pi_n^J(0)}^{\text{PDV of Continuation Cost}}$$

$$= \frac{(1 - e^{-rT_n})(h_n c_n + f_n)}{r} + \sum\nolimits_{i=i_n^0,\ldots,I} e^{-rA_{in}}\theta_i + e^{-rT_n}\pi + e^{-rT_n}\Pi_n^J(0).$$

Notice that

$$\Pi_n^J(0)$$

takes on a nice closed form expression, which is given by

$$\Pi_n^J(0) = \int_0^{L_n} e^{-rt}\left[h_n c_n^J + f_n^J\right] t\,dt + \sum_{i=1,..,I} e^{-r(\alpha_i/(h_n s_n))}\theta_i + e^{-rL_n}\pi + e^{-rL_n}\Pi_n^J(0)$$

$$= \frac{h_n c_n^J + f_n^J}{r} + \frac{\sum_{i=1,..,I} e^{-r(\alpha_i/(h_n s_n))}\theta_i + e^{-rL_n}\pi}{1-e^{-rL_n}}.$$

The life-cycle cost of operating the entire fleet of aircrafts at $t = 0$ is therefore given by

$$\Omega(E_1^0, E_2^0, \ldots, E_N^0) = \sum_n \Pi_n(E_n^0).$$

Having described how we calculated the life-cycle costs of operating the fleet, we can now calculate the marginal cost per FH of a one-time flight at $t = 0$. This cost depends on whether the flight is home based or deployed and varies across aircraft according to the following calculation:

$$\text{One-Time Cost/FH} = \begin{cases} \underbrace{c_n}_{\text{Direct Cost}} + \overbrace{\Pi_n(E_n^0 + s_n^H) - \Pi_n(E_n^0)}^{\text{Change in Lifecycle Cost}} \text{ if home based;} \\ \underbrace{c_n}_{\text{Direct Cost}} + \overbrace{\Pi_n(E_n^0 + \omega_n s_n^H) - \Pi_n(E_n^0)}^{\text{Change in Lifecycle Cost}} \text{ if deployed.} \end{cases}$$

Estimates of the Full Marginal Cost of Additional C-130 Flying Hours

This section provides estimates of the average marginal cost by command. We break the marginal cost estimates up into the direct cost incurred at $t = 0$ and changes in costs that accrue over time on a present discounted value. In addition to describing the cost effects, we also characterize the marginal change in the date of aircraft replacement associated with each case.

Table C.6 summarizes the findings when we used OMB's 2.7 percent annual real discount rate. These findings suggest that an additional FH costs, on average, $5,900 in one-time direct costs and $900 in future costs on a discounted cash flow basis. The $900 in future costs represents the present discount value of speeding up major maintenance activities and replacement, as well as the savings that may be due to replacing an older C-130 variant sooner with the newer C-130J model, which tends to cost less to operate. The fleetwide average effect does not vary much depending on whether the one-time additional FH is while the aircraft is deployed or home stationed; however, differences do exist across commands.

In particular, the cost of an additional FH on future discounted costs is, on average, approximately zero at AMC and greater than the fleetwide average at ANG. The fact that AMC's future discounted costs are near zero deserves explanation. As discussed earlier in this appendix, replacing a C-130E and C-130H aircraft with a C-130J at AMC leads to significantly greater operating cost savings. In turn, speeding up the time to aircraft replacement yields nontrivial cost savings. Offsetting these benefits are, of course, the costs associated with performing major maintenance and procuring new aircraft sooner. In AMC's case, these two factors approximately cancel each other out.

Table C.6
Summary of Marginal Cost Estimates and Changes in Time to Aircraft Replacement by Command at an Assumed Annual Real Discount Rate of 2.7 Percent

	One-Time Cost ($000)	Change in Future Costs ($000)	Full Marginal Cost ($000)	Change in Time to First Replacement (days)
	a	b	c = a + b	
One-hour flight at *t* = 0 for a home-based aircraft				
AFRC	5.9	1.2	7.1	–0.9
AMC	5.8	0.0	5.8	–0.5
ANG	5.9	1.2	7.0	–1.0
PACAF	5.9	0.9	6.8	–0.5
USAFE	5.8	0.9	6.7	–0.6
Overall average	5.9	0.9	6.8	–0.8
One-hour flight at *t* = 0 for a deployed aircraft				
AFRC	5.9	0.8	6.6	–0.5
AMC	5.8	–0.1	5.8	–0.6
ANG	5.9	1.4	7.3	–1.1
PACAF	5.9	1.1	7.0	–0.6
USAFE	5.8	1.1	6.9	–0.7
Overall average	5.9	0.9	6.8	–0.8

Table C.7 considers a higher real discount rate of 5.0 percent per year. Under the higher discount rate, the discounted future net costs of a one-time additional FH actually increase to $1,500 in the home-stationed case and $1,600 in the deployed FH case, from the $900 level found under the lower discount rate of 2.7 percent. This may seem counterintuitive because a greater discount rate generally decreases net present value. We obtained our result because the future costs associated with major maintenance activities and replacement occur earlier than the benefits associated with faster movement to the less-expensive-to-operate C-130J aircraft. This means that the costs from aging the aircraft fleet tend to be frontloaded relative to the operating cost benefits of switching to a C-130J at an earlier date. A higher discount rate will tend to reduce the present value of the benefits that accrue from the lower operating costs associated with the C-130J. Overall, the discount rate sensitivity analysis suggests that the choice of discount rate is important and can materially affect the estimates.

Table C.7
Summary of Marginal Cost Estimates and Changes in Time to Aircraft Replacement by Command at an Assumed Annual Real Discount Rate of 5.0 Percent

	One-Time Cost ($000)	Change in Future Costs ($000)	Full Marginal Cost ($000)	Change in Time to First Replacement (days)
	a	b	c = a + b	
One-hour flight at $t = 0$ for a home-based aircraft				
AFRC	5.9	1.5	7.4	–0.9
AMC	5.8	1.0	6.9	–0.5
ANG	5.9	1.6	7.5	–1.0
PACAF	5.9	2.0	7.9	–0.5
USAFE	5.8	2.1	7.9	–0.6
Overall average	5.9	1.5	7.4	–0.8
One-hour flight at $t = 0$ for a deployed aircraft				
AFRC	5.9	0.9	6.8	–0.5
AMC	5.8	1.2	7.1	–0.6
ANG	5.9	1.9	7.8	–1.1
PACAF	5.9	2.4	8.3	–0.6
USAFE	5.8	2.6	8.4	–0.7
Overall average	5.9	1.6	7.4	–0.8

APPENDIX D

The Commercial Intratheater Airlift Optimization Model

This appendix outlines a mathematical formulation for the cost minimization of CITA movements in the USCENTCOM AOR. The model accounts for organic airlift and for two modes of commercial airlift. These civilian systems include the contractual leasing of commercial aircraft and the tendering of items via TEP. In terms of payload, the model tracks the movements of both cargo and passengers. In this particular AOR, commercial airlift does not handle passenger traffic for the U.S. military. Consequently, organic lift is the only available means of moving personnel.

The CITA model itself is a large-scale MILP, which uses integer assignments for such factors as the assignment of aircraft to routes and the choice of whether to tender via a TEP carrier. The model's objective is to minimize the total cost of airlift over the modeled time horizon, and these costs include the operating costs for military assets, funds spent to acquire leased aircraft, and the cost to hire tender carriers. The operational optimization platform is an eight-core workstation running Cplex, a commercial optimization solver. Cplex is able to multithread a computer's individual cores, such that each CPU can process a distinct node on the branch-and-bound tree to expedite solution times. In the most computationally challenging scenarios we examined, the model required approximately one week of clock time (or 50 days of CPU time) to complete.

Modeling Structure, Elements, and Data

Before entering a detailed description of the CITA model, it is important to first present its fundamental indices. These elements, which will become subsets in the algebraic formulation, are

- *ac* = aircraft, drawn from an organic pool of C-130s and C-17s, and IL-76s as contract assets
- *cargo* = type of payload, either cargo or passengers
- *b* = bases within USCENTCOM where cargo is available for either pickup or delivery
- *r* = routes, which are predefined as an ordered set of bases, *b*
- *t* = time, which ranges from day 1 to day 365 of calendar year 2009.

The largest data pool feeding the CITA model was the set of cargo requirements in USCENTCOM during the calendar year 2009. As discussed in Appendix A, military and contract movements came from the GATES database, TEP movements from a spreadsheet

USTRANSCOM's Acquisition Directorate supplied us. During this time frame in USCENTCOM, airlift handled approximately 1.1 million passenger movements and 400,000 tons of cargo.

In the model, each organic and contracted aircraft may fly along one of approximately 1,000 prescribed routes on each scenario day. These routes were not necessarily those flown in 2009, but each route was created from sorties that were flown by at least one organic cargo asset in that time frame. The model disallows a particular aircraft from one of these routes if it would be unable to complete the route implementing quick turn ground times within a 16-hour duty day. TEP aircraft, however, were not assigned to any specific route in the model. During the bidding process, a bid awardee agrees to haul the tender from its point of embarkation to its point of debarkation in a method most convenient to the carrier. Delivery, however, must occur in a time frame agreed on prior to award of the tender.

The model has a special methodology for employing t, the time index. While it is theoretically possible to optimize cargo movements over the entirety of 2009 at one time, this is unrealistic for two key reasons. First, it implies that, from day 1, planners have perfect knowledge of all future cargo requirements and could schedule cargo pickups and deliveries accordingly. Perfect visibility of all information in advance is highly unlikely and is not representative of historical experience in the AOR. Second, the size of an optimization matrix that manages cargo over the full time horizon would be extremely large, and the difficulty of solving the resultant integer problem would be immense. Even a capable solution platform would be hard pressed to solve to even an approximate optimum in a reasonable time frame.

In practice, planners typically have knowledge of upcoming cargo movements with roughly 24 to 48 hours' notice. The CITA model implements this degree of advance information by employing a rolling time window whose width is two days (see Figure D.1). During optimization, the model has knowledge of which cargo is available to move "today." Call this initial period t_1. The model also has visibility of cargo that will become available to move "tomorrow" (period t_2). As shown in the figure, the model then determines the most cost-effective allocation of cargo to aircraft and aircraft to routes within this two-day window. Next, it stores the optimal movements for period t_1 and then advances the time horizon by 24 hours. The MILP now has access to a near-feasible starting point for period t_2. This advanced basis can accelerate the solution for the new window, which now extends from t_2 to t_3. The model re-solves, stores optimal cargo allocations and routings, and again advances the time counter. The solution iteration proceeds until the conclusion of the scenario year.

Model Constraints

The constraints in this airlift system can be described as falling into one of four categories: aircraft management, cargo management, TEP bid mechanics, and the objective function. These categories are described in more detail below.

Aircraft Management

There are two important aspects for tracking aircraft from day to day in the scenario. First, the number of aircraft assigned to each route on each day, $ASSIGN_{ac,r,t}$, should sum to the available inventory of that aircraft type on that day, $NUMAC_{ac,t}$:

Figure D.1
Mechanics of the Sliding Time Window

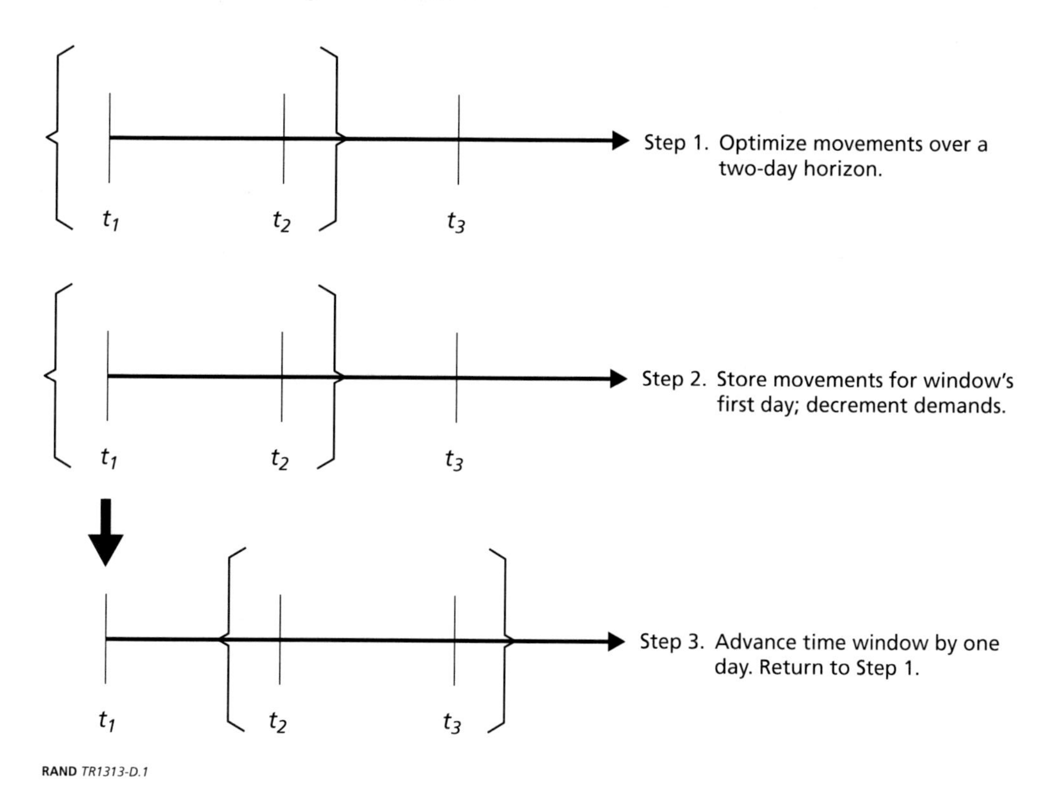

$$\sum_{r} ASSIGN_{ac,r,t} = NUMAC_{ac,t} \forall ac,t.$$

Note that, in this chapter's formulation, elements in all lower-case indicate sets. Terms in all caps represent system variables; items in mixed case represent the model's input data or constant values. For example, in this equation, we have introduced two variables, *ASSIGN* and *NUMAC*. Both *ASSIGN* and *NUMAC* are integer variables representing a discrete count of aircraft. During execution of the model, *NUMAC* is held constant for C-130s for the entire time horizon in each scenario. The number of C-17s is a constant for each quarter of 2009, and the value is derived from the average of the number of tails observed in USCENTCOM in the historical record.

The number of IL-76s available, however, is variable and controlled entirely by the optimization routine. Because contracts for leased aircraft typically last for one year, the number of contract aircraft may increase only monotonically during the time frame modeled here. Consequently, a floor is put in place for *NUMAC,* as shown in equation D.2 based on the number of contract aircraft leased during the previous scenario day. The model may then add new IL-76s to the leased fleet in each new window to represent the augmentation of the contract fleet:

$$NUMAC_{'IL-76',t} \leq NUMAC_{'IL-76',t+1} \forall t.$$

It is possible to lease an arbitrarily large fleet of contract aircraft early in the year and to allow unneeded airframes to sit idle until a future spike in demand requires their use. Such behavior is generally not cost-effective because contract language typically specifies a penalty for the nonuse of a leased airframe. This nonuse fee helps the carrier recoup the revenue lost by not leasing the aircraft to another client. Planners in 2009 were very sensitive to this particular penalty and generally had lease arrangements for fewer than a half-dozen aircraft at any time. We will further address cost issues with the contract fleet in the subsection on to the minimum cost objective function.

Second, we required the endpoint of the route an aircraft flies on a scenario day to be the origin of the route for the next day's flight. Aircraft may thus not reposition overnight. In the model's route structure, aircraft could remain overnight in only eight locations in a total of six countries: Bagram and Kandahar, Afghanistan; Ali Al Salem, Kuwait; Thumrait, Oman; Balad and Al Sahra, Iraq; Al Udeid, Qatar; and Manas, Kyrgyzstan. In a practical sense, this emulates a small set of midsize to large regional bases that possess the security and maintenance sufficient for efficiently protecting and servicing a significant subset of the mobility fleet.

Accordingly, we introduced two logical sets, *head* and *tail*:

$$\sum_{r \in tail(r,b)} ASSIGN_{ac,r,t} = \sum_{r \in head(r,b)} ASSIGN_{ac,r,t+1} \forall ac,t,r,b.$$

The membership of these sets is limited to the bases that are the respective start or conclusion of the route in question.

Cargo Management

The principal aspects of delivering cargo in the model are the assurances that no cargo should be picked up before its available-to-load date (ALD) and that it should be delivered no later than its required delivery date (RDD):

$$\begin{aligned}
&\sum_{t' \leq t;\ cargo \neq 'pax'} TENDERED_{b,b',t'} + \\
&\sum_{r;t' \leq t;\ cargo,\ ac \in okcargo(cargo,ac)} TRANSPORT_{r,b,b',ac,t'} \geq \\
&Demand_{b,b',cargo,t} \forall b,b',cargo,t.
\end{aligned}$$

Two new continuous variables appear in this relation. *TENDERED* represents the amount of cargo moved by TEP, and these movements may not include passengers. The *TRANSPORT* variable represents the amount of cargo moving by either organic or contract means. Contract carriers are prohibited from carrying passengers, and the pairing of *ac* = IL76 with *cargo* = pax is explicitly blocked from membership in the logical set, *okcargo*.

The *Demand* parameter is also introduced here. *Demand* is computed prior to optimization and takes into account the cumulative cargo demands on each scenario day based on the ALD and RDD of each historical movement. An item's ALD was assumed to be the cargo pickup date drawn from GATES. This is a conservative assumption, because cargo may well have been available for pickup prior to its historical loading date. Another conservative assumption is that a cargo's modeled RDD occurs on the actual date of its delivery, as shown in the GATES database. Where GATES did not provide a delivery time, the RDD was assumed to be five days after the ALD.

The CITA model also maintains an accurate mass balance of cargo awaiting delivery from one day to the next within the scenario. This mass balance requires a new continuous variable, *CARGOREM*, which tracks the amount of cargo that is available to move from its origin to its destination, but has not yet been picked up for delivery:

$$CARGOREM_{b,b',cargo,t} = CARGOREM_{b,b',cargo,t-1} + Demand_{b,b',cargo,t}$$
$$-TENDERED_{b,b',t}\delta_{cargo=bulk}$$
$$-\sum_{r;cargo,ac \in okcargo(cargo,ac)} TRANSPORT_{r,b,b',ac,t}, \forall b,b',cargo,t.$$

In short, the cargo remaining for delivery today accounts for what remained yesterday, what became available today, and what moved today. One should note that, when the RDD is greater than the ALD, there is no penalty for the early delivery of cargo. In fact, there are often cost efficiencies to be realized by loading a single plane with two pallets at once rather than delivering one pallet today and the second tomorrow.

Finally, an aircraft's load should exceed its capacity at no point along its route. As many of the model's routes consist of more than a single onload and a single offload point, we required a more-sophisticated logical set, *loadok*, to specify which loads on a route will contribute to the aircraft's capacity constraint during any sortie on each route. Consider the notional four node route shown in Figure D.2. The valid pairings of b and b' in *loadok* are denoted in the figure by directed arcs. In this example, cargo may move from b_1 to b_2, but not vice versa. The arcs that overlap vertically are those whose loads would contribute simultaneously toward the aircraft's capacity constraint. So as that aircraft moves from b_1 to b_2, the total load must account not only for cargo moving between the two bases but also for the base pairs b_1-b_3 and b_1-b_4.

The algebraic representation for tracking load capacity in the model is:

$$\sum_{cargo=bulk;b,b' \in loadok(r,b,b')} TRANSPORT_{r,b,b',ac,cargo,t}$$
$$+\sum_{cargo=pax;b,b' \in loadok(r,b,b')} PaxScale_{ac} TRANSPORT_{r,b,b',ac,cargo,t}$$
$$\leq MaxLoad_{ac} ASSIGN_{ac,r,t} \forall ac,r,t.$$

MaxLoad is a parameter representing the maximum cargo payload of each aircraft. Our modeling convention assumes that aircraft payloads are limited on the basis of weight, rather than volume. To reflect the fact that mobility aircraft are often volume- rather than weight-limited,

Figure D.2
Valid Linkages in *Loadok* for a Four-Node Route

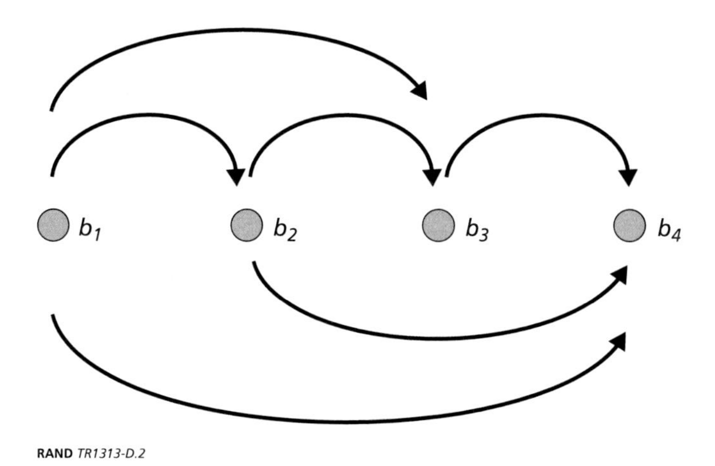

RAND *TR1313-D.2*

we used the average load factors presented in AF Pamphlet 10-1403 as maximum allowable payloads; these average load factors are considerably less than an aircraft's maximum allowable weight, to reflect the effects of volumetric constraints on achievable payloads. It is possible to load both cargo and passengers onto the same aircraft.[1] When this occurs, the model assumes a linear trade-off between the aircraft's maximum cargo tonnage and its maximum passenger capacity. For example, the model assumes that the cargo capacity for a C-17 is 45 tons. A C-17 can also hold a maximum of 188 passengers, which is possible with palletized seating kits. If the model opts to carry, say, 89 passengers (one-half its capacity), it can also carry 22.5 tons of cargo (one-half its maximum cargo load). This linear trade-off between cargo and passengers appears in the constraint via the parameter *PaxScale*.[2]

The CITA model lacks a precise volumetric, or "cube" constraint to limit the cubic feet and/or the number of cargo pallet positions to the volumetric capacity of each aircraft. While there is a field in GATES for pallet positions, we observed that the historical records corresponding to USCENTCOM movements in 2009 were largely unreliable, and thus we did not implement a cube constraint in the current formulation. The AMC planning factors for load limit (as published in AFPM 10-1403) that the model used are generally quite conservative for the short sorties most common in the USCENTCOM AOR. For example, recall the planning factor mentioned earlier for the maximum load on a C-17, 45 tons. The C-17 can carry a maximum of 85 tons, so this notional maximum is used during planning to account for frequent loading of less-dense cargo that may take up available space on the aircraft before it reaches the maximum payload weight. With such conservative planning factors in play, especially in conjunction with relatively short sorties, activation of a cube constraint would be unlikely to influence the model's outcomes.

TEP Bid Mechanics

After conducting regression analyses of the history of successful bids in the TEP program, we found three variables to be instrumental in determining the price of moving cargo. First, the distance between the cargo's origin and its destination plays a significant role. This seems quite intuitive; flying longer distances requires more fuel and increases operations and maintenance expenses for load-carrying aircraft. Second, heavier tenders were found to cost more than lighter loads. Again, this seems reasonable because it takes more fuel to move the cargo than it does to fly an aircraft empty. Third, a frequent history of bids along an origin-destination pair implies a smaller expected tender cost. This is the least sensitive of the three variables, but it stems from a carrier's willingness to bid lower to obtain regular tenders on a more-frequent, reliable transportation arc. Figure D.3 depicts a notional piecewise linear cost profile for tendered cargo, emphasizing the relative importance of cargo weight, tender distance, and tender frequency.[3]

Careful incorporation of these cost factors into an optimization model is important to avoid nonlinearities and concave features in formulating a convex minimization problem. To this end, we introduced the concept of weight and history "bins" to simplify these relevant

---

[1] In our historical 2009 data set, 38 percent of intra-USCENTCOM C-17 sorties carried both cargo and passengers, and 23 percent of C-130 sorties carried both cargo and passengers.

[2] This is equivalent to assuming that each passenger consumes a fixed portion of the aircraft's maximum payload weight.

[3] The x-intercept is meaningless here, since a tender of zero tons is nonsensical.

Figure D.3
Notional Tender Cost Profile

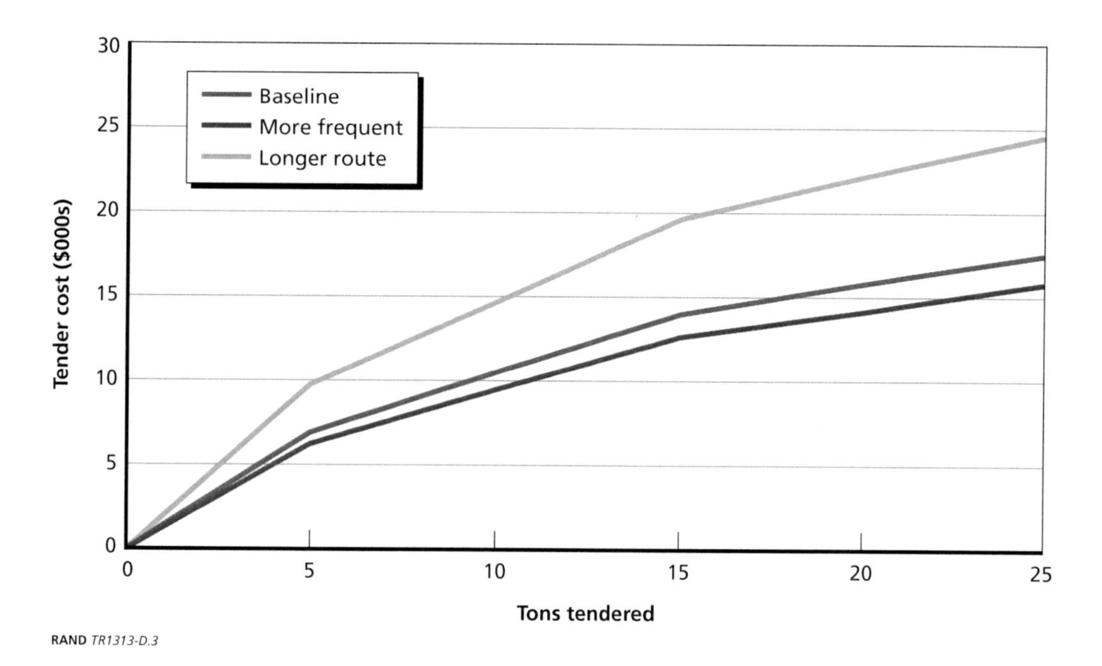

RAND *TR1313-D.3*

TEP bid mechanics. Cargo is treated as fitting into one of three bin sizes, according to the weight of the tender: up to 5 tons, more than 5 tons and up to 15 tons, and more than 15 tons but less than 25 tons.[4] On a given scenario day, a tender will also fit into one of three bins based on the history of movements along its origin-destination pair: less than 20 bids on this arc thus far this year, between 20 and 40 bids thus far, and more than 40 bids to date.

To accommodate this, we introduced a binary variable, $BID_{b,b',wt,hist}$. *BID* is indexed over the now-familiar sets of origin base, *b*; destination base, *b*'; and scenario day, *t*. It is also indexed over two new sets, which link to two new, related parameters:

- *wt* = weight bin; 1, 2, or 3
- *hist* = history bin; 1, 2, or 3
- $Weight_{wt}$ = weight ceiling for bin *wt*
- $History_{hist}$ = floor for history bin *hist*.

BID allowed us to select a single weight bin, and the optimization will view only a single (and locally convex) cost in that bin. We assigned a single cost to each bin, derived from that bin's cost at its weight ceiling. Figure D.4 depicts the transformation of the baseline cost profile from Figure D.3. The transformation approach has one primary advantage: The resultant objective

---

[4] The use of three bins may appear to be a simple approximation to the tender cost profile. Higher resolution in the approximation of the cost function would be preferable, but such resolution would come at the cost of an increase in the number of binary elements in the *BID* variable. In general, the computation time necessary to solve a MILP grows polynomially with the number of integer and binary variables. Because the binary BID represents the majority of discrete variables in the model, increasing the resolution from three to four tender cost bins would represent roughly a 4/3, or 133 percent, growth in the count of discrete variables. Such a growth in variable count could be expected to lead to a solution time growth of at least $(4/3)^2$, or 80 percent longer to solve the model, which is already quite long (approximately 50 CPU days) in the base case using three bins.

Figure D.4
Transformation of Notional Cost Profile

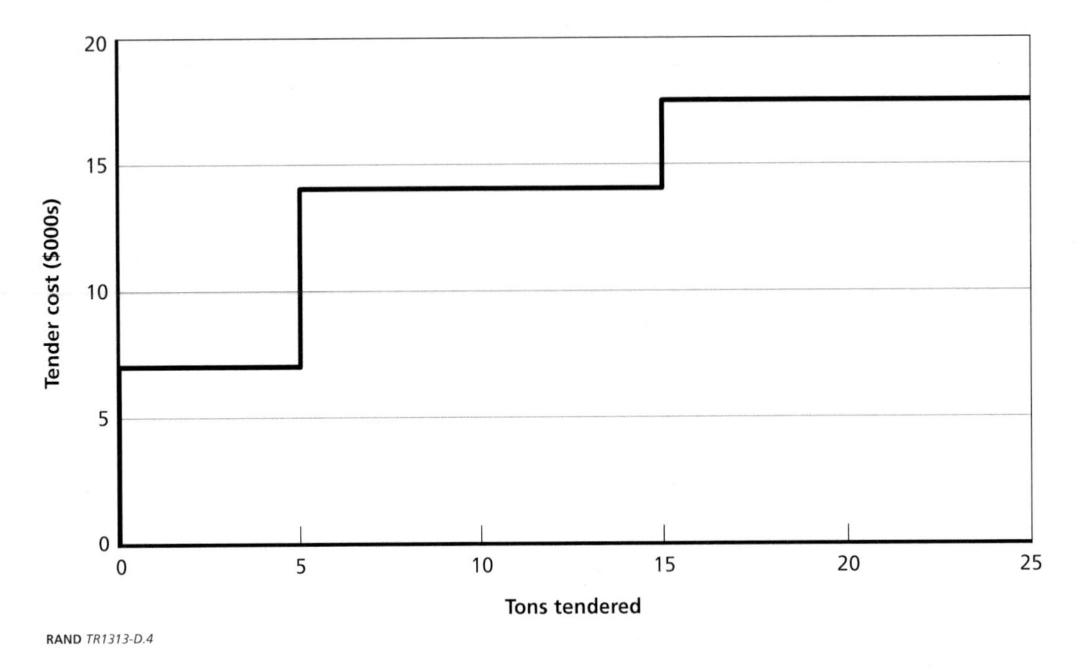

RAND *TR1313-D.4*

function is convex. In optimizing TEP movements, the model tends to assign the bulk of movements within a bin close to its weight ceiling. This is sensible behavior and in accordance with the cost profile because the most cost-effective expenditure per ton moved occurs at the bin break points. The model is fairly insensitive to a change in these break points. In a sensitivity analysis, the break point vector for the bin ceilings was shifted from (5, 15, 25) to (2.5, 7.5, 25) tons, but the two objective values were indistinguishable within the convergence criterion established for an integer solution.

Capturing the mechanics of the bid process required three fundamental constraints. First, no more than one bid may be accepted for each origin-destination pair on any given day:

$$\sum\nolimits_{wt,hist} BID_{b,b',t,wt,hist} \leq 1 \forall b,b',t.$$

Second, tendered cargo must fit into one of the three weight classes:

$$TENDERED_{b,b',t} \leq \sum\nolimits_{wt,hist} Weight_{wt} BID_{b,b't,wt,hist} \forall b,b',t.$$

Considered alone, this constraint might allow a single tender to fall into multiple weight classes. For example, a tender of 12 tons fits under the ceiling for weight bins 2 (up to 15 tons) and 3 (up to 25 tons). However, in a minimum-cost environment, the optimization will select the lower weight class to realize the maximum cost savings.

Third, the history of tendered cargo along each origin-destination pair must fit into one of the history bins:

$$\sum\nolimits_{t'<t,wt,hist} BID_{b,b',t',wt,hist} \geq \sum\nolimits_{wt,hist} Hist_{hist} BID_{b,b',t,wt,hist} \forall b,b',t.$$

As with the weight bin restriction, the history bin constraint might allow a single tender to fall into multiple history bin classes. The minimum-cost objective, however, encourages the optimization to select the history bin class with the highest floor to effect the most significant cost savings.

Objective Function

Finally, the model's cost function and its continuous objective variable, *COST*, are given by

$$COST = \sum_{ac,r,t} RouteCost_{ac,r} ASSIGN_{ac,r,t} + \sum_{b,b',t,wt,hist} BidCost_{b,b',wt,hist} BID_{b,b',t',wt,hist}.$$

Two new parameters appear in the objective, $RouteCost_{ac,r}$ and $BidCost_{b,b',wt,hist}$. *RouteCost* denotes the cost for either organic or contract lift to fly on a particular route. Organic aircraft incur a flat cost per flying hour, whereas charter costs were based on USTRANSCOM's FY 2010 contract with Silk Way Airlines for intra-USCENTCOM airlift. The contract language generally specifies a flat rate between particular airfields or a per mile charge for origins and destinations not specifically stated in the contract. The contract also specifies a penalty for not using a leased airframe, as discussed earlier in the aircraft management constraints. *BidCost* represents the price for TEP airlift based on the regression analysis' key variables of origin-destination pair, weight class, and history bin.

Bibliography

Air Force Doctrine Document (AFDD) 2-6, *Air Mobility Operations*, March 1, 2006.

Air Force Instruction (AFI) 65-503, *U.S. Air Force Cost and Planning Factors*, various appendixes, various dates.

Air Force Pamphlet (AFPAM) 10-1403, *Air Mobility Planning Factors*, December 18, 2003.

Air Mobility Command, "Deployed versus Home Station Usage," spreadsheet, undated.

———, "USAF C-130 TAI 16 Jul 09," spreadsheet, undated.

Anderson, Don, "Tender Program," briefing, Air Mobility Command, Directorate of Analyses, Assessments and Lessons Learned, undated.

———, Air Mobility Command, Directorate of Analyses, Assessments and Lessons Learned, email to the authors, June 7, 2011.

Baker, Steven F., David P. Morton, Richard E. Rosenthal, and Laura Melody Williams, "Optimizing Military Airlift," *Operations Research*, Vol. 50, No. 4, 2002, pp. 582–602.

Cutter, David, AMC/A9, personal communication via email, October 5, 2012.

DoD—*See* U.S. Department of Defense.

Huard, Dean A., "The Theater Express Program: A Combat Logistics Force Multiplier," *Army Sustainment*, July–August 2011, pp. 38–41.

Joint Publication (JP) 3-17, *Air Mobility Operations*, October 2, 2009.

Newton, Clayton T., and Dennis Turnage, *The Defense Distribution Center's Future Role in Theater Distribution Operations*, graduate thesis, U.S. Army War College, March 2007.

Office of Management and Budget, "Real Treasury Interest Rates," 2009. As of February 16, 2012: http://www.whitehouse.gov/omb/circulars_a094_a94_appx-c/

Omdall, Christopher N., *Air Cargo Tenders: Theater Express for the World*, graduate thesis, Air Force Institute of Technology, June 2010.

Orletsky, David T., Daniel M. Norton, Anthony D. Rosello, William Stanley, Michael Kennedy, Michael Boito, Brian G. Chow, and Yool Kim, *Intratheater Airlift Functional Solution Analysis (FSA)*, Santa Monica, Calif.: RAND Corporation, MG-818-AF, 2011. As of December 4, 2012: http://www.rand.org/pubs/monographs/MG818.html

Skyline Aviation Ltd., "Ilyushin IL-76," undated. As of February 20, 2012: http://www.skylineaviation.co.uk/downloads/IlyushinIL-76.pdf

Snyder, Donald, and Patrick Mills, *Supporting Air and Space Expeditionary Forces: A Methodology for Determining Air Force Deployment Requirements*, Santa Monica, Calif.: RAND Corporation, MG-176-AF, 2004. As of February 20, 2012: http://www.rand.org/pubs/monographs/MG176.html

Therrien, Kevin C., "Theater Airlift: An Analysis of Star Routes vs. Optimized Scheduling," graduate thesis, Air Force Institute of Technology, June 2003.

U.S. Air Force, *FY 2013 Budget Overview*, February 2012. As of February 16, 2012:
http://www.saffm.hq.af.mil/shared/media/document/AFD-120209-052.pdf

U.S. Code, Title 10, Section 2640, Charter Air Transportation of Members of the Armed Forces, 2011.

———, Title 49, Section 40118, Government-Financed Air Transportation ("Fly America Act"), 2011.

———, Title 49, Section 41106, Airlift Service ("Fly CRAF Act"), 2011.

U.S. Department of Defense, *Mobility Capabilities and Requirements Study 2016 (MCRS-16): Executive Summary*, February 2010. As of February 16, 2012:
http://www.airforce-magazine.com/SiteCollectionDocuments/TheDocumentFile/Mobility/MCRS-16_execsummary.pdf

U.S. Transportation Command, "Letter to Valued AMC TWCF Customers," August 17, 2010.

———, "Contract Terms and Conditions," spreadsheet, detailing Contract Number HTC711-09-D-5004-P00003, undated.